Measuring Alcohol Consumption

Measuring Alcohol Consumption

Psychosocial and Biochemical Methods

edited by

Raye Z. Litten and John P. Allen

National Institute on Alcohol Abuse and Alcoholism,

Rockville, Maryland

Humana Press **Totowa, NJ**

This book is based on a workshop sponsored by the National Institute on Alcohol Abuse and Alcoholism (NIAAA)

Library of Congress Cataloging-in-Publication Data

Main entry under title:

Measuring alcohol consumption : psychosocial and biochemical methods /
        edited by Raye Z. Litten and John P. Allen.
                        p. cm.
                "Based on a workshop sponsored by the National Institute on
        Alcohol Abuse and Alcoholism (NIAAA)"--T.p. verso.
                Includes index.
                ISBN 0-89603-231-0
                        1. Alcoholism—Diagnosis—Research—Methodology—Congresses.
        2. Drinking of alcoholic beverages—Research—Methodology—
        Congress. 3. Alcoholism—Physiological aspects—Congresses.
        I. Litten, Raye Z. II. Allen, John P. III. National Institute on
        Alcohol Abuse and Alcoholism (U.S.)
                        [DNLM: 1. Alcohol Drinking—congresses. 2. Data Collection—
        methods—congresses.   WM 274 M484]
        RC565.M3458 1992
        616.86'1—dc20
        DNLM/DLC
        for Library of Congress   92-1536
                                                                        CIP

Preface

The Importance of Measuring Alcohol Consumption

To date, alcohol studies have attended far more to issues of alcohol dependence and the harmful consequences of drinking than to the level of alcohol consumption itself. This is, perhaps, not surprising since dependence on alcohol is believed to constitute a meaningful and distinct medical syndrome, regardless of the level of alcohol consumption associated with it (Edwards and Gross, 1976). Also, society is generally more concerned with the adverse consequences of drinking (e.g., traffic fatalities, homelessness, health care and legal expenses, and academic/behavioral problems in young drinkers), than with the quantity of alcohol actually consumed.

Nevertheless, accurate assessment of alcohol usage is important in its own right in at least four contexts:

1. Evaluating the effectiveness of alcoholism and alcohol abuse treatment and prevention efforts. Such efforts include both applied evaluations of existing programs and formal, well-controlled efficacy studies on experimental interventions. These investigations require rigorous methodologies to assess outcomes precisely and contrast what may be quite subtle differences between programs and between pretreatment and posttreatment outcomes. Although these studies are usually characterized by the employment of multiple measures of success—including general improvements in social and physical functioning, reduction in degree of dependence, and resolution of problems directly resultant from drinking—it is by their assessment of changes in drinking behavior that they are potentially able to achieve the highest level of objectivity and exactitude. The stated goal of most alcoholism treatment programs in the US is total abstinence, a condition that can only be validated by exact quantification of alcohol consumption. For those programs that aim at moderated consumption,

it is even more important precisely to measure both pretreatment drinking and posttreatment levels of consumptuion over time so that the actual impact of the intervention in achieving its stated goal can be demonstrated. Though the evaluation of prevention and treatment programs must include more than assessment of subsequent consumption, other outcome variables, such as changes in attitudes, health status, dependence, and problems ascribed to drinking, are by their very nature more susceptible to measurement error in that they rely more heavily on a subject's interpretation and expression of internal states than does level of consumption. So too, changes in alcohol consumption likely correlate with these other outcomes and may predict them.

2. Assessment of the level of alcohol consumption is probably the single most important component of screening for future alcohol-related behavioral and medical problems. Other screening indices, be they levels of hepatic enzymes or verbal admission of alcohol-related difficulties, can be accurately interpreted only if one knows the subject's level of drinking. To date, biochemical screening markers, in particular, have tended to be nonspecific to alcoholism since they are influenced by a variety of other medical and lifestyle conditions. Commonly used self-report screening instruments, such as the Michigan Alcoholism Screening Test, focus primarily on the consequences of alcohol use and on the symptoms of alcoholism. Since they rarely impose a time frame for the questions, recovering alcoholics even with years of uninterrupted sobriety or individuals with distant histories of problematic drinking may be expected to continue to score well within the clinical range. Thus, it is important to appreciate the significance of other biological and self-report screening items that will allow the examiner to ascertain the respondent's current level of alcohol consumption. Interestingly, carefully worded questions about the level of consumption might also prove less threatening and more specific and, hence, be more likely to be answered accurately than typical verbal screening items dealing with others' reactions to one's drinking, embarrassment, and guilt over drinking.

3. Alcohol consumption, with or without collateral alcoholism, is itself an important risk factor for a multiplicity of serious medical problems (Babor, Kranzler, and Lauerman, 1987) from fetal alcohol effects, which seem primarily precipitated by acute peak maternal blood alcohol levels during critical periods of prenatal development (Clarren, Bowden, and Astley, 1987) to conditions such as coronary artery disease (Altura,

1986), pancreatitis (Van Thiel et al., 1981), and various liver pathologies (Grant, Dufour, and Harford, 1988), the risks for all of which seem to be exacerbated with increasing levels of consumption. A quick, inexpensive, and accurate means of assessing alcohol consumption would greatly assist primary care physicians in diagnosing medical problems caused or aggravated by heavy alcohol use and in advising patients on how their drinking may be associated with future physical problems.

4. Finally, monitoring alcohol use itself is important in assuring public safety. Society maintains an interest in total abstinence by airline pilots, air traffic controllers, railway engineers, and medical personnel on duty. Similarly, individuals charged with alcohol-related traffic offenses may be ordered by the Court to refrain completely from drinking for some period of time as a condition of probation or maintenance of a restricted driving license. Assurance of abstinence obviously requires an accurate means of measuring consumption.

To focus attention on methods for measuring alcohol consumption, the National Institute on Alcohol Abuse and Alcoholism (NIAAA) convened a working conference of leading experts. The meeting was held April 4 and 5, 1991 in Bethesda, MD. The techniques discussed encompassed a broad range of approaches from direct verbal and collateral accounts of drinking to complex and innovative biochemical indicators; from techniques that summate drinking over an extended period of time to procedures that yield a measure of relatively recent consumption. Not included in the discussion were methods such as the analysis of blood, breath, or urine that provide only brief and immediate measures of drinking. Each speaker reviewed the status of a particular technique, described its application in both clinical and research settings, and specified what research remains still necessary to refine the procedure or make it more practicable. Those presentations, appropriately enhanced by the fruits of the discussion, comprise the body of this text.

The second day of the Conference allowed deliberation on comparisons and contrasts of techniques, broad methodological issues surrounding applications of the procedures, and strategies for incorporating the techniques into a range of settings. Salient aspects of their discussion serve as the foundation for the summary chapter of this text.

Beyond providing investigators a forum in which to discuss their research and its implications with each other, the Conference was valuable for NIAAA staff. It allowed proponents of the various techniques to recommend directions that the Institute might pursue in recommending methodologies for assessing alcohol consumption in treatment and research contexts. This book represents the first formal effort to integrate and disseminate this information to the broader range of clinical practitioners and alcohol researchers.

John P. Allen
Raye Z. Litten

References

B. M. Altura (1986) Introduction to the symposium and overview. *Alcoholism (NY)* **10,** 557–559.

T. F. Babor, H. R. Kranzler, and R. J. Lauerman (1987) Social drinking as a health and psychosocial risk factor: Anstie's limit revisited. *Recent Developments in Alcoholism* (M. Galanter, ed.), Plenum, NY, **5,** 373–402.

S. K. Clarren, D. M. Bowden, and S. L. Astley (1987) Pregnancy outcomes after weekly oral administration of ethanol during gestation in the pig-tailed macaque (Macaca nemestrina). *Teratology* **35,** 345–354.

G. Edwards and M. M. Gross (1976) Alcohol dependence: Provisional description of a clinical syndrome. *Br. Med. J.* **1,** 1058–1061.

B. F. Grant, M. C. Dufour, and T. C. Hartford (1988) Epidemiology of alcoholic liver disease. *Seminars in Liver Disease* **8,** 12–25.

D. H. Van Thiel, H. D. Lipshitz, L. E. Porter, R. R. Shade, G. P. Gottlieb, and T. O. Graham (1981) Gastrointestinal and hepatic manifestations of chronic alcoholism. *Gastroenterology* **81,** 594–615.

Contents

Contributors

John P. Allen • *National Institute on Alcohol Abuse and Alcoholism, Rockville, MD*

Raymond F. Anton • *Medical University of South Carolina, Charleston, SC*

Thomas F. Babor • *University of Connecticut School of Medicine, Farmington, CT*

Olof Beck • *Karolinska Hospital, Stockholm, Sweden*

Stefan Borg • *St. Görans's Hospital, Stockholm, Sweden*

Gerald J. Connors • *Research Institute on Alcoholism, Buffalo, NY*

Paul Cushman, Jr. • *Northport Veterans Administration Hospital Center, Northport, NY*

Frances K. Del Boca • *University of Connecticut School of Medicine, Farmington, CT*

Douglas R. Gavin • *Addiction Research Foundation of Ontario, Toronto, Ontario, Canada*

Margo S. George • *University of Ottawa, Ottawa, Canada*

Anters Helander • *St. Görans's Hospital, Stockholm, Sweden*

Charles S. Lieber • *Veterans Affairs Medical Center, Bronx, NY*

Renee C. Lin • *Indiana University, School of Medicine, Indianapolis, IN*

Raye Z. Litten • *National Institute on Alcohol Abuse and Alcoholism, Rockville, MD*

Lawrence Lumeng • *Indiana University, School of Medicine, Indianapolis, IN*

Stephen A. Maisto • *Brown University Medical School, Providence, RI*

Michael Phillips • *St. Vincent's Medical Center of Richmond, Staten Island, NY*

Alan S. Rosman • *Veterans Affairs Medical Center, Bronx, NY*
Harvey A. Skinner • *University of Toronto, Toronto, Ontario, Canada*
Linda Sobell • *University of Toronto, Toronto, Ontario, Canada*
Mark B. Sobell • *University of Toronto, Toronto, Ontario, Canada*
Helena Stibler • *Karolinska Hospital, Stockholm, Sweden*
Larry Swette • *Giner, Inc., Waltham, MA*
Robert M. Swift • *Brown University, Providence, RI*
Annetta Voltaire • *St. Görans's Hospital, Stockholm, Sweden*

PSYCHOSOCIAL MEASURES OF ALCOHOL CONSUMPTION

Just the Facts

Enhancing Measurement of Alcohol Consumption Using Self-Report Methods

Thomas F. Babor and Frances K. Del Boca

Introduction

"Evidence.... is the life of Truth, and Method the life of Discourse; the former being requisite to convince the understanding, the latter, to facilitate the searches of it."[1]

During the 1960s, one of the most popular American television programs was called *Dragnet*, a police melodrama that clearly portrayed the forensic methods used in the pursuit of criminals, truth, and justice. To Sergeant Joe Friday, the indefatigable police detective, evidence was the life of truth, and the self-report method was the way to "facilitate the searches of it." Friday knew how to get information from suspects and witnesses by persistent probing, a professional attitude, and a polite way of returning the respondent to "just the facts."

This article deals with the collection of facts about alcohol consumption by means of the self-report method, a procedure that requires a respondent to answer questions about recent or past use of alcoholic beverages either verbally or in writing. After providing a brief history of the self-report method and the different applications it has served, a heuristic model of the question-answering process is

From: *Measuring Alcohol Consumption*
Eds.: R. Litten and J. Allen ©1992 The Humana Press Inc.

reviewed. This model is designed to provide a conceptual basis for recommendations about how to improve the reliability and validity of verbal report methods. By focusing on the way in which questions are asked and responded to, some logical inferences are made about the sources of unreliability and invalidity. Finally, this paper concludes with a discussion of how the reliability and validity of self-report information can be improved to the point where it is useful for various types of applied research such as treatment outcome investigations, epidemiological studies, population surveys, and studies of the treatment process.

Applications of Self-Report Methods

The self-report method has become the dominant means of collecting information about alcohol use. This is true not only for scientific research but also for clinical applications and marketing surveys. In the area of scientific research, self-report methods have been used to estimate the prevalence of alcohol use and abuse in population surveys, to study the consequences of alcohol consumption in epidemiological investigations, and to evaluate the effectiveness of treatment in clinical studies.[2-4] In other applications, the self-report method has been used to screen for hazardous alcohol consumption in public health programs and to collect information about drinking preferences in marketing surveys for the alcoholic beverage and insurance industries.[5,6] Given its ease of use and relatively low cost, it is not surprising that the self-report method has become the dominant procedure for collecting information about alcohol consumption.[3,7]

Reliability and Validity

The concepts of reliability and validity have been applied to estimate two types of error associated with the use of self-report measures in different study populations.[8] The first, random error, is caused by uncontrollable and unpredictable random influences. This type of error affects the reliability of a response. Reliability is defined as the consistency of an individual's reporting both within a single assessment session and between two occasions. The second type of error, called systematic or fixed bias, occurs when something distorts a response every time a measurement is made. This kind of bias results in either a systematic overestimation or underestimation of the person's alcohol consumption. Reliability is a necessary but not sufficient condition for validity. This implies that one important source of invalidity is unreliability. As a consequence, this paper will describe methods for estimating and enhancing reliability as well as approaches for reducing systematic response bias or invalidity.

Since self-report methods have become more popular as a means of measuring alcohol consumption, increasing attention has been devoted to questions about their reliability and validity. Studies by Fuller et al. (1988), Edwards (1985), Watson et al. (1984), and Pendery et al. (1982) have suggested that the validity of data obtained from persons with alcohol-use disorders is suspect.[9–12] This skepticism has its origins in the earliest formulations of Alcoholics Anonymous, which identified denial of problem drinking as the hallmark of alcoholism.[13] The assumed existence of denial has logically led to skepticism regarding self-report data.

Recent reviews of methodological studies have concluded that although information obtained from alcoholics generally tends to be reliable and valid, there can be considerable variability in accuracy depending on many factors:[3,14–16]

1. The sensitivity of the information sought (e.g., demographic data vs arrest records)
2. The specificity of the validation criteria (e.g., archival records, breath alcohol readings, urine tests, informant's reports)
3. The personal characteristics of the respondents (e.g., sober vs intoxicated)
4. The time window of the report (e.g., lifetime vs recent) and
5. The demand characteristics of the task situation (e.g., clinical interview vs research evaluation).

These reviews indicate that verbal report data pertaining to alcohol consumption are inherently neither valid nor invalid but vary with the methodological approach of the data gatherer and the personal characteristics of the respondent. This conclusion suggests that the question of whether self-report drinking data are either valid or invalid is inappropriate. The more important question is *what conditions are necessary to produce reliable and valid self-report information for a given purpose.*

The Question-Answering Process

In an effort to provide a framework for understanding the optimal conditions for obtaining self-report information, Babor and colleagues have proposed a conceptual model that borrows from the principles of learning, memory, cognition, motivation, and social influence.[3,17] As shown in Fig. 1, this model specifies the major sources of unreliability and invalidity in the question-answering process. This formulation is similar to a model of response bias proposed by Cannell, Miller, and Oksenberg (1981) and builds on research on cognitive and social psychology, eyewitness testimony, and survey methods.[18–21] After a brief description of the model, evidence will be reviewed to suggest how each stage of the question-answering process is susceptible to response error. More importantly, the model will also be used to suggest how response error can be reduced, eliminated, or controlled.

As depicted in Fig. 1, the question-answering process occurs in a general social context that is governed by prevailing cultural norms and the constraints of the immediate interpersonal situation, which may include other respondents, research personnel, or treatment staff. Within this social context, the respondent's self-report is directly influenced by a combination of task variables, respondent characteristics, motivation, and cognitive processes.

Task variables provide a context for the question-answering process by defining the purpose of the assessment (e.g., scientific research vs triage to treatment). As such, they can have an important influence on the respondent's motivation to respond accurately. Similarly, the degree of rapport with the interviewer, the way in which a questionnaire is presented, the extent of confidentiality implied in the procedure, and the likelihood of future verification all place demands on the respondent's veracity. Other task variables include the complexity and duration of the task and the time interval considered (e.g., last week vs last year).

One situation that can adversely affect the validity of clinical information is when data are collected by the treating clinician. The desire by the respondent to be viewed as successfully treated may lead to underreporting in outcome assessments. Conversely, clients responding to questions at the beginning of treatment may overreport consumption to justify their need for treatment.[22]

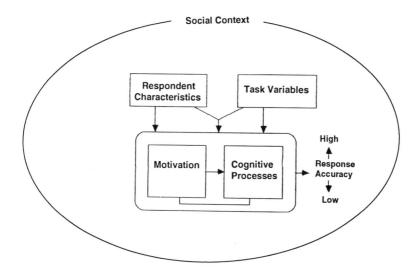

Fig. 1. Schematic diagram of the question answering process that leads to either accurate or inaccurate responses. Reprinted with permission from *Behavioral Assessment* 12, Thomas F. Babor, Joseph Brown, and Frances K. Del Boca, Validity of Self-Reports in Applied Research on Addictive Behaviors: Fact or Fiction?, 1990, Pergamon Press, Inc.

Related to task variables are a variety of factors that can influence the reliability of measurement. When alcohol consumption is assessed by means of an interview, the skill of the interviewer becomes an important factor in determining response accuracy. This skill is reflected in asking questions, probing answers, coding responses, and recording responses. Consistency across interviews conducted by the same person and across different interviewers will help to reduce random error. Key entry of data, an activity that occurs outside the parameters of this model, represents another way in which error can be introduced; procedures that maximize accuracy should be established.

Respondent characteristics are enduring qualities such as personality traits, attitudes, beliefs, intelligence level, and cognitive impairment, as well as more transitory conditions such as state of sobriety with regard to alcohol or other drugs, physical condition (e.g., fatigue or withdrawal), and psychological state (e.g., depression).

Research on response bias has shown that heavy drinkers tend to underestimate alcohol use when they have a positive blood-alcohol level.[3] Other research has shown that with continued posttreatment sobriety, there is general improvement in cognitive function, personality adustment, and health status.[23,24] These factors are likely to affect an individual's ability to comprehend questions and recall information.

Another factor thought to affect the validity of self-reports is "need for approval," the desire to present oneself in a favorable fashion. In clinical studies that question patients about their drinking, the social desirability response set is thought to result in an underestimate of alcohol use, although there is no compelling evidence to support this.[25]

A critical variable in response accuracy that is related to both respondent characteristics and situational factors is motivation. The more respondents are motivated to conform to the interviewer's or investigator's instructions and expectations, the more likely they will be to give accurate information.

Motivation is most likely to be affected by the degree of threat or embarrassment associated with a question or line of inquiry. It is often assumed that alcohol consumption is seen by the general public as undesirable and therefore may be underreported. Respondents may be motivated to minimize, modify, or deny their drinking because they want to present themselves in a favorable light. This social desirability hypothesis assumes that all respondents—not just those who are high in need for approval—will find questions about drinking behavior threatening. However, alcoholics in treatment may be quite willing to talk about their drinking and may even be motivated to overreport their use in order to resolve the cognitive dissonance associated with seeking treatment.[22]

Another source of error associated with motivation is denial. Within the popular lore of Alcoholics Anonymous, denial is thought to distort the accuracy of information about heavy drinking. Denial may be related to the degree of social undesirability of a given problem or behavior. If so, it is more likely to affect answers to questions that imply personal defects, such as drinking problems, than to questions about normative behaviors, such as drinking. Within the context of the present model, denial is conceived as the outcome (inaccurate responding) of a complex, social-psychological process rather than its cause.

As shown in Fig. 1, a specific self-report is the result of an interplay of motivational and cognitive factors. Cognitive processes thought to influence the respondent's ability to answer questions about drinking include attention, comprehension, retrieval, integration, and response selection. The respondent must attend to a request for information (attention), interpret the question (comprehension), recall from memory pertinent drinking behaviors (retrieval), integrate information via comparative, inferential, or attributional processes (integration), and then decide how to respond (response selection). This set of interconnected processes is influenced by the inherent cognitive limitations of the respondent, which are most likely to impair his or her ability to remember past events accurately. Retrievel is strongly affected by the recency and saliency of past events.[20] Memory error can also result from cognitive impairment, memory decay (forgetting), response distortion (e.g., repression), and telescoping (i.e., events remembered as occurring more recently). If the respondent is cognitively impaired by chronic drinking, intoxicated at the time of the assessment, or too depressed to think clearly, retrieval may be affected.

Cognitive processing is influenced by motivational factors such as the respondent's short-term goals (e.g., avoiding arrest or obtaining treatment), level of arousal, and affective state. At low arousal levels, the respondent is not likely to attend to requests for information. In the final step (integration), the respondent evaluates the psychological meaning of the response in relation to personal goals, such as self-esteem or fear of being labeled an alcoholic. If the potential response is evaluated as non-threatening, the respondent provides the correct response. If the personal goals of the respondent are inconsistent with an accurate response (e.g., to appear abstinent to the program staff), then the response is more likely to be modified.

This hypothetical model of the question-answering process illustrates the complex environmental, social, and cognitive demands placed on respondents who answer questions about alcohol consumption. On a practical level, the model suggests ways in which response bias can be minimized. As discussed elsewhere in more detail, these procedures include increasing the respondent's motivation, facilitating the retrieval process, clarifying comprehension of the question, and structuring the task so that accurate responding is perceived as the primary task.[3,8,17] The next two sections summarize procedures that have been suggested to minimize response bias by enhancing reliability and validity.[8,21,26,27]

Enhancing the Reliability of Self-Report

There are several ways to improve the reliability of interview assessments. Detailed manuals provide instructions that assure consistency in scripting, questioning, probing, and coding of responses. Staff training should be intensive and should provide general principles of methodology and interviewing technique, as well as specific instruction for particular instruments. Training should include observation of model interviews and practice sessions with actual clients, followed by performance feedback. Once a performance criterion is reached, procedures for routine monitoring of interviews should be in place to assure that quality is maintained. There are several specific exercises that can be used in training and in quality-assurance monitoring to establish and maintain interviewer reliability.

Tape-Rating Task

This task involves taping an interview as a model for interviewing technique and as a means of assessing consistency in coding responses. Interviewers are asked to observe the tape and complete the interview protocol as the tape progresses. Observers' entries are compared to the standard, discrepancies are noted, and additional instruction is given to improve coding accuracy in future interviews.

Interviewer–Observer Exercise

A representative of the client population is recruited. One staff member interviews the client and records responses. A second staff person observes the interview and as in the tape-rating exercise, also completes the interview form. In this way, interviewers can observe each other, provide feedback on performance, and develop a consistent approach to dealing with situations that arise in the course of administering the interview to actual clients.

Test–Retest Exercise

Respondents are interviewed on two separate occasions by two different interviewers. Discrepancies in recorded responses are noted, and possible reasons for inconsistencies are discussed with the respondent. Unlike the first two exercises, the test–retest procedure allows an assessment of interviewer reliablity; that is, actual interviewer performance can be compared, and alternative approaches to handling the same client can be discussed.

Scripted Respondents

This exercise involves the development of a drinking history that is codified as a set of responses to a particular instrument. The completed form is then used as a standard for evaluating interviewer reliability. Staff are directed to conduct the interview with an "actor" whose task is to answer questions in accordance with the established drinking history. The actor can participate in as many interviews as needed. If the actor is properly trained, his or her responses, unlike those of actual respondents used in a test-retest exercise, should not alter as a function of repeated assessments. In addition to training, these exercises can be used to estimate the reliability coefficients for evaluating the adequacy of particular interviewers and specific instruments.

Enhancing the Validity of Self-Reports

Based on the multidimensional model described in Fig. 1, the validity of self-report information can be improved by procedures that provide a better definition of the respondent's task, improve the respondent's motivation, and facilitate cognitive processing of desired information.

Defining the Task

A crucial aspect of the task is the specification of the respondent's role (e.g., research subject or patient). The performance requirements (e.g., accurate responding) also need to be clearly stated. Often, instructions are not sufficiently clear concerning what is expected of the respondent. Procedures designed to reduce invalidity resulting from ambiguous task demands are specific instructions about the purpose of the questions, modeling of response requirements by means of clear examples, and instructional passages before difficult questions. Instructions that encourage the respondent to work hard, organize answers, and report accurately even if the answer is embarrassing should increase motivation and improve accuracy. The most common techniques employed in collecting alcohol consumption data are standard instructions concerning anonymity and confidentiality. Researchers have generally avoided mixing clinical and research functions in the person of the same interviewer by employing independent research assistants to collect treatment outcome data. Other task characteristics thought to influence accurate responding are follow-up (probe) questions, clarity of language, question length, and wording. Longer questions seem to encourage higher response rates by giving the respondent more time to think.[18,28]

Improving Motivation

Procedures recommended to increase the patient's motivation to respond accurately include: a commitment agreement, clear instructions, periodic reinforcement of role behavior, the "bogus pipeline" method, and embedding questions in a less threatening context.

The commitment agreement involves asking the respondent to make an overt (signed) agreement to work hard to provide complete and accurate information.[18] Reinforcement of role behavior can be accomplished by prompting the respondent to "take time to think carefully" and periodically thanking the subect for good performance. Self-report assessment can be compared to alternative sources of information to improve motivation. Respondents whose reports of abstinence contradict biochemical data or reports of significant others may be able to explain discrepancies.

"Bogus pipeline" is another motivational technique that has been employed to increase accurate responding by drinkers.[29–31] In this procedure, subjects provide information under conditions where they are led to believe that objective, external validation of their response is available to the data gatherer. The possibility that inaccuracies can be detected is hypothesized to increase the validity of self-report data.

Several "pipelines" can be used to indicate to the patient that researchers are capable of validating their true responses to questions about drinking. For example, laboratory test results (e.g., liver enzyme tests) that may be sensitive to recent and chronic heavy drinking can be performed after a blood sample is taken to corroborate estimates provided by the respondent. One study found that almost twice as many pregnant women reported consuming alcohol when they were told their verbal reports would be verified by means of blood and urine tests.[31]

Improving Cognitive Processing

Memory errors can be minimized by use of memory aids (e.g., timeline follow-back procedure), better instructional passages, longer questions, aided recall (e.g., fixed-response choices), and bounded recall (i.e., presenting results of previous responses that may be a reminder for present response). Recall procedures using memory aids provide cues to the respondent as part of the interview. A good example is the timeline follow-back (TLFB) technique, which presents respondents with a calendar and asks them to recall their drinking on a daily basis.[14]

Another technique is to make the questions as specific as possible. Information tends to be more accurate when the drinking behavior is referenced to an exact time period rather than to what is regular or usual. Global questions using words such as usually, generally, and regularly may be subject to misunderstanding and response distortion.[32]

The quality of reporting may be improved by adding memory cues, especially when they increase the length of the question. Longer questions give the respondent more time to think and recall important events or behaviors. Because the length of a reply is generally proportional to the length of a question, longer questions may generate more information.

Drinking diaries are another procedure that reduce reliance on recall. Self-report diary entries collected prospectively are often referred to as self-monitoring.[33] One potential confounding effect associated with keeping a diary is reactance, i.e., the tendency for respondents to change their behavior as a function of focusing their attention on it. Self-monitoring reduces reliance on recall and may therefore be especially suitable when studying populations expected to have cognitive deficits that may impair recall.

Conclusions

In general, self-report procedures provide reasonably accurate estimates of alcohol consumption, but this statement needs to be qualified. Response accuracy depends on a variety of conditions, such as the wording of questions, the time frame being remembered, the motivation of the respondent, the respondent's state of sobriety, and the mode of administering the questions. The more these conditions are optimal, the more accurate the response. Accuracy also needs to be considered in terms of the purpose of the data-gathering exercise and the availability of alternative methods of assessment. Self-report methods, when properly applied, can produce valid information on nominal, ordinal, and interval scales, especially relative to biological markers and observational methods.

Self-report methods have become the most widely used procedures for estimating alcohol consumption because of their flexibility, simplicity, and inexpensiveness. Questionnaires are the most economical means of gathering drinking data, followed by computer-administered interviews, telephone interviews, and face-to-face interviews. Biochemical tests and other procedures may be less expensive than interviews because of training and personnel costs.

Although self-report procedures tend to be less accurate for past drinking than for current drinking, they have been used to reconstruct an individual's drinking history from adolescence through adulthood. When precision is required, a time window of one or two months is optimal for self-report methods, which are subject to the vagaries of human memory. When precision is not required, retrospective accuracy may be more than adequate to distinguish between global categories of alcohol consumption such as heavy, moderate, or light drinking—or abstinence—during a longer period.

Practitioners tend to appreciate the usefulness of verbal report procedures, since medical histories rely primarily on this method. The problem is that practitioners do not generally like to ask structured questions, which are necessary for reliable and valid information. Clients rarely refuse to respond to questionnaires or interviews dealing with drinking.

The unique benefits of self-report procedures are their flexibility, adaptability, inexpensiveness, and efficiency. They are highly portable and can be linked to the respondent through a variety of communications technologies, such as the telephone, computer, and even television. Although social science training is important for the design and administration of these methods, they do relatively well when employed by persons having a minimum of technical training.

Recommendations for Further Research

The variety of verbal report procedures used to assess drinking behavior makes it difficult to evaluate and compare results across studies. Until recently, little attention has been focused on the systematic evaluation of measurement procedures, assessment contexts, and response sets. Alcohol research would benefit from a new generation of methodological studies directed at factors that enhance the validity of verbal report data and evaluate new procedures that provide objective indicators of recent alcohol consumption.

A recent study based on a model of the question-answering process presented above provides an example of this type of research.[34] This investigation was conducted solely to evaluate the effects of two respondent factors (alcohol status at the time of data collection and level of cognitive function) and one task variable (method: personal interview vs self-administered questionnaire) on the reliability and validity of self-reported substance use. In addition, the test–retest reliability of the questionnaire was evaluated. Self-reports of alcohol, marijuana, and cocaine use were validated by means of toxicologic analyses of blood and urine samples collected in the same time interval. Overall, self-reports were found to be reliable and valid. In terms of respondent variables, recent alcohol consumption resulted in lower validity coefficients, whereas level of cognitive functioning had no impact on response accuracy.

The generic term for factors that might influence verbal report information in a research context is response effects. While there has been some research in this area, little attention has been devoted to alcohol use. There are several opportunities for further research in this area. The following recommendations, summarized from recent literature reviews, may serve as useful guidelines for a future research agenda.[3,35]

In recent years, there has been a proliferation of reliability studies that compare one estimation method to another or the same method on two different occasions. Most of this research has been methodologically weak and lacks a coherent theoretical approach. Rarely have reliability and validity studies been designed exclusively to investigate specific questions. Rather, the research to date has been primarily derived from secondary analysis of data collected for other purposes. There is a need to move beyond "add-on" studies and secondary analyses to a more systematic approach to research on the question-answering process.

More studies of assessment contexts and instructional sets are needed. For instance, studies suggest that individual assessments yield more valid results than group assessments, that telephone surveying and interviewing may be a reliable means of obtaining alcohol consumption information, that adolescents will give more valid self-reports in private settings, and that the use of computerized assessments may yield valid data as do other techniques. This kind of research should be encouraged.

Individual difference factors (e.g., organic impairment, forgetting, and social desirability response sets) need to be investigated because they may potentially influence validity. But until it can be demonstrated that social desirability scales actually measure response bias on alcohol-related assessments, their use should not be automatically incorporated into assessment batteries.[36]

Since no single measure of alcohol consumption is entirely valid, most researchers recommend the use of convergent lines of evidence to establish drinking status or measure drinking behavior. Analytic procedures such as the multitrait-multimethod assessment matrix should be employed to evaluate the discriminant and convergent validity of verbal report data. Different verbal report strategies for quantifying consumption (e.g., quantity–frequency questions, diary method, and TLFB) should be compared. The use of multiple biological indicators may have promise as a way of corroborating verbal reports, provided sensitivity problems can be overcome.

In summary, verbal data obtained from alcoholic and nonalcoholic respondents are inherently neither valid nor invalid but vary with the methodological sophistication of the data gatherer and the personal characteristics of the respondent. This suggests that emphasis on demonstrating whether verbal report data are either accurate or inaccurate is misplaced. A more important issue is what conditions are conducive to response accuracy and what procedures contribute to valid responses. An approach to response distortion that takes into account the social psychological nature of the question-answering process provides an alternative to the psychodynamic view of denial that rejects verbal report data as essentially unreliable. Good research practice dictates that the best way to deal with response bias is to focus on reliability- and validity-enhancement procedures. The approach outlined in this paper could be used to design a program of research that systematically varies constituent components of the question-answering process to determine the extent to which respondent characteristics, task variables, motivational constructs, and cognitive processes contribute to the accuracy of information provided by the verbal report method.

References

[1]Anonymous (1735) *The Art and Mystery of VINTNERS and Wine-Coopers: or a Brief Discourse Concerning the Various Sicknesses and Corruptions of Wines* (J. Clarke, London), p. 1.

[2]M. E. Hilton (1987) Drinking patterns and drinking problems in 1984: Results from a general population survey. *Alcoholism* **11,** 167–175.

[3]T. F. Babor, R. S. Stephens, and G. A. Marlatt (1987) Verbal report methods in clinical research on alcoholism: Response bias and its minimization. *J. Stud. Alcohol* **48,** 410–424.

[4]H. R. Kranzler, T. F. Babor, and R. J. Lauerman (1990) Problems associated with average alcohol consumption and frequency of intoxication in a medical population. *Alcoholism: Clin. Exp. Res.* **14(1),** 119–126.

[5]J. B. Saunders, O. G. Aasland, A. Amundsen, and M. Grant (in press) W.H.O. collaborative project on early detection of persons with harmful alcohol consumption: Alcohol consumption and related problems among primary health care patients. *Brit. J. Addict.*

[6]G. Armyr, A. Elmer, and U. Herz (1982) *Alcohol in the World of the 80s: Habits, Attitudes, Preventive Policies and Voluntary Efforts* (Sober Forlags Ab., Stockholm).

[7]L. C. Sobell and M. B. Sobell (1986) Can we do without self-reports? *Behav. Therapist* **9,** 141–146.

[8]H. A. Skinner (1981) Assessment of alcohol problems: Basic principles, critical issues and future trends, *Research Advances in Alcohol and Drug Problems.* Y. Israel, F. B. Glaser, H. Kalant, R. E. Popham, W. Schmidt, and R. G. Smart, eds. (Plenum, New York), pp. 319–369.

[9]R. K. Fuller, K. K. Lee, and E. Gordis (1988) Validity of self-report in alcoholism research: Results of a veterans administration cooperative study. *Alcoholism: Clin. Exp. Res.* **12,** 201–205.

[10]G. Edwards (1985) A later follow-up of a classic case series: D. L. Davies's 1962 report and its significance for the present. *J. Stud. Alcohol* **46**, 181–190.

[11]C. G. Watson, C. Tilleskjor, E. A. Hoodecheck-Schow, J. Pucel, and L. Jacobs (1984) Do alcoholics give valid self-reports? *J. Stud. Alcohol* **45**, 344–348.

[12]M. L. Pendery, I. M. Maltzman, and L. J. West (1982) Controlled drinking by alcoholics?: New findings and a reevaluation of a major affirmative study. *Science* **217**, 169–175.

[13]R. E. Tarter, A. I. Alterman, and K. L. Edwards (1984) Alcoholic denial: A biological interpretation. *J. Stud. Alcohol* **45**, 214–218.

[14]L. C. Sobell, S. A. Maisto, M. B. Sobell, and A. M. Cooper (1979) Reliability of alcohol abusers' self-reports of drinking behavior. *Behav. Res. Ther.* **17**, 157–160.

[15]L. T. Midanik (1988) The validity of self-reported alcohol use: A literature review and assessment. *Brit. J. Addict.* **83**, 1019–1029.

[16]T. J. O'Farrell and S. A. Maisto (1987) The utility of self-report and biological measures of alcohol consumption in alcoholism treatment outcome studies. *Adv. Behav. Res. Ther.* **9**, 91–125.

[17]T. F. Babor, J. Brown, and F. K. Del Boca (1990) Validity of self-reports in applied research on addictive behaviors: Fact or fiction? *Behav. Assess.* **12**, 5–31.

[18]C. F. Cannell, P. V. Miller, and L. Oksenberg (1981) Research on interviewing techniques, *Sociological Methodology*. J. Leinhardt, ed. (Jossey-Bass, San Francisco), pp. 389–437.

[19]S. J. Sherman, C. M. Judd, and B. Park (1989) Social cognition, *Annual Review of Psychology*, vol.40, M. R. Rosenweig and L. W. Porter, eds. (Annual Review, Palo Alto, CA), pp. 281–326.

[20]E. F. Loftus (1979) *Eyewitness Testimony* (Harvard University Press, Cambridge, MA).

[21]N. M. Bradburn and S. Sudman, eds. (1981) *Improving Interview Method and Questionnaire Design* (Jossey-Bass, San Francisco).

[22]L. Midanik (1982) Over-reports of recent alcohol consumption in a clinical population: A validity study. *Drug Alcohol Depend.* **9**, 101–110.

[23]T. F. Babor, Z. Dolinsky, B. Rounsaville, and J. Jaffe (1988) Unitary versus multidimensional models of alcoholism treatment outcome: An empirical study. *J. Stud. Alcohol* **49**, 167–177.

[24]J. T. Becker and J. H. Jaffe (1984) Impaired memory for treatment-relevant information in inpatient men alcoholics. *J. Stud. Alcohol* **45**, 339–343.

[25]N. M. Bradburn (1983) Response effects, *Handbook of Survey Research*. P. E. Rossi and J. D. Wright, eds. (Academic, New York), pp. 289–328.

[26]C. F. Cannell and R. L. Kahn (1968) Interviewing, *The Handbook of Social Psychology*, vol.2, 2nd Ed., G. Lindzey and E. Aronson, eds. (Addison-Wesley, Reading, MA), pp. 526–595.

[27]A. M. Cooper, M. B. Sobell, L. C. Sobell, and S. A. Maisto (1981) Validity of alcoholics' self-reports: Duration data. *Intl. J. Addict.* **16**, 401–406.

[28]N. M. Bradburn, S. Sudman, and Associates (1979) *Improving Interview*

Method and Questionnaire Design: Response Effects to Threatening Questions in Survey Research. Jossey-Bass, San Francisco.

[29]E. E. Jones and H. Sigall (1970) The bogus pipeline: A new paradigm for measuring affect and attitude. *Psychol. Bull.* **76,** 349–364.

[30]E. M. Botvin, G. J. Botvin, N. L. Renick, A. D. Filazzola, and J. P. Allegrante (1984) Adolescents' self-reports of tobacco, alcohol, and marijuana use: Examining the comparability of videotape, cartoon and verbal bogus-pipeline procedures. *Psycholog. Reports* **55,** 379–386.

[31]J. B. Lowe, R. A. Windsor, B. Adams, J. Morris, and Y. Reese (1986) Use of a bogus pipeline method to increase accuracy of self-reported alcohol consumption among pregnant women. *J. Stud. Alcohol* **47,** 173–175.

[32]W. A. Belson (1981) *The Design and Understanding of Survey Questions* (Gower, Aldershot, UK).

[33]R. E. Vuchinich J. A. Tucker, and L. M. Harrlee (1988) Behavioral assessment, *Assess. Addict. Behav.* D. M. Donovan and G. A. Marlatt, eds. (Guilford Press, New York), pp. 51–83.

[34]J. Brown, H. R. Kranzler, and F. K. Del Boca (in press) Self-reports of alcohol and drug abuse inpatients: Factors affecting reliability and validity. *Brit. J. Addict.*

[35]Institute of Medicine (1989) *Prevention and Treatment of Alcohol Problems: Research Opportunities* (National Academy Press, Washington, DC).

[36]D. P. Crowne and D. Marlowe (1960) A new scale of social desirability independent of psychopathology. *J. Consult. Psychol.* **24,** 349–354.

Computerized Approaches to Alcohol Assessment

Douglas R. Gavin, Harvey A. Skinner, and Margo S. George

Introduction

Improving the accuracy of alcohol assessment remains a perplexing challenge for researchers and clinicians. Computer technology has attracted growing attention as a possible solution. Rapid developments in computer hardware and software offer considerable yet little-explored potential for improving the reliability and validity of assessments. Indeed, there is a popular conception that individuals may be more honest when responding to sensitive questions (e.g., alcohol use) to a computer than in a face-to-face interview. Moreover, computers have a mystique that can be useful for engaging hard-to-reach populations.[1]

Computer-aided testing has a history that extends over three decades. Slack et al. (1966) showed that medical patients could operate a computer, allowing them to complete a medical history without any other aid.[2] However, the feasibility of testing patients was limited in practical terms because of the complicated computer language and large expense of using a mainframe computer. As technology progressed, however, the cost of computers reduced dramatically—by 20–30% a year. The year 1975 marked the appearance of the first

From: *Measuring Alcohol Consumption*
Eds.: R. Litten and J. Allen ©1992 The Humana Press Inc.

"build-it-yourself" microcomputers that were affordable and easily programmable. Home-computer hobbyists were the first to adopt this new technology for more individual use. Shortly thereafter, the microcomputer revolution took off when the mass-produced Apple II (1977) and IBM PC (1981) microcomputers allowed the average person access to computer technology. During this period, computer-aided testing quickly became a realistic alternative to traditional interviews and pencil-and-paper tests.

This chapter has three related aims. First, an overview is given of the advantages of using computer technology for assessments. Second, various studies are described that have evaluated what effect computerization has on the reliability and validity of alcohol assessments, and third, new research is presented on an innovative use of computers, i.e., measuring response latency, for increasing the accuracy of alcohol use.

Why Use Computers?

Several reasons are offered for the use of microcomputers, each pertaining to a different aspect of the assessment process.[3] The most frequently given reasons for computer-aided testing are speed, efficiency, and accuracy of testing. Other reasons include the generation of computer-aided diagnosis, instant feedback concerning responses, and the mystique of computer use.[3,4]

Speed and Efficiency

One problem in particular often confronts busy clinicians: How can I more effectively use my time? Carr, Ghosh, and Ancill (1983) suggest that taking a full medical or psychiatric history is a formidable procedure, often involving some 200 questions.[5] They argue that the patient may feel that many of the items have little relevance to current problems and would much rather spend their limited time with the clinician discussing more pressing concerns. Much of the background information on clients could easily be gathered by a computer, leaving more time available for direct patient–doctor interaction. Computer-aided testing can also provide an instant scoring service for the test.

Accuracy

A second, primary advantage of computerized testing relates to the accuracy of scores. Testing accuracy comes from several sources, and computerized testing offers advantages in each domain. These domains consist of procedural error that can occur in filling out questionnaires and the motivational set with which the test taker approaches the assessment situation.

Procedurally, computerized testing routinely prevents specific types of errors that can occur in interviews or pencil-and-paper tests. For example, no hand coding or scoring is required. This eliminates data entry problems, addition mistakes, and questions ambiguously circled or skipped. Irrelevant sections can be skipped automatically, resulting in a less-confusing flow to the assessment.

In terms of psychological set, answering questions one-by-one on a computer may be less overwhelming to the patient than being faced with a stack of typed pages. The computer can present each question by itself, subtly suggesting that it rates special attention. These points will likely result in better cooperation and attention from the test taker.

There is also some evidence that people may respond more honestly to questions presented by a computer. Even and Miller (1969) argue that computerized testing increases the perception that responses are kept confidential, increasing the client's comfort in giving honest responses.[6] Further, interviews done face-to-face make the social desirability of responses more salient, and nonverbal cues could provide unintentional feedback from the interviewer. The impact of social desirability is especially relevant to alcohol assessment, given the social stigma frequently attached to excessive drinking levels and symptoms of alcohol dependence.

Tailored Testing

Another advantage results from the computer's ability to score items as the test is being carried out. Instantaneous scoring allows "dynamic" testing to take place: The test adapts to the respondent's previous answers. This process has been referred to as tailored, adaptive, response-contingent, or branched testing.[7] The overall advantage of this process lies in a shorter test with the same degree of accuracy. Each question is scored after it is asked, and the respondent's final score is estimated. The next question is asked to refine the estimate, until a final score is established. In alcohol-dependence assessment, for example, responses to initial items would be used to generate a rough estimate of the level of severity, whereupon additional items would be administered from that domain of severity in order to fine tune the estimate. Examples of tailored testing have been successful in saving 50–75% of the test time required.[8,9] Aside from the obvious time savings, quicker testing also leads to better acceptance of computer testing by patients and practitioners.

Unobtrusive Measures

Computers also offer some advantages not available with pencil-and-paper tests or face-to-face interviews: The client's test-taking behavior can be measured without his or her being conscious of the measurement taking place. For example, some sections of a questionnaire may call for additional questions because of one of the client's responses. By having the computer automatically change the questionnaire, the branching is not readily apparent. No hints are given that extra or different questions are being used or that sections of the questionnaire have been skipped. Further, any branching is handled automatically and is therefore not subject to human error.

In addition, the computer can measure the amount of time the subject takes to complete each question. This response latency gives information about how reliably and accurately the subject is answering each question. In particular, overly short response times are associated with inaccurate responses, and overly long times are indicative of inconsistent ones.[10] This issue will be addressed in detail in the third section of this chapter.

Instant Feedback

In addition to more accurate test data entry and scoring, the computer can also provide instant feedback as well as aid diagnosis. These qualities can improve assessment in several ways: The use of color graphics can hold attention throughout the testing process, making respondents more likely to cooperate, to respond honestly, and to be willing to take additional tests later. In making the test process more enjoyable, the increased attention should also result in more accurate answers. For personal feedback, the computer can provide instant analysis and a quick summary of the client's answers. This can be provided directly and interactively to the client, in which case it may have more impact than delayed feedback. This feedback would also make the testing session more interesting and challenging.

The computer provides rapid feedback to the clinician, who then has more information available to make immediate decisions. The computer could score the data more comprehensively and offer "expert advice" concerning different types of profiles found in the data. In short, test results are available in less time and with the potential for more detail. A final printed report is often readily available as part of the test scoring process.

Acceptance of Computers

Although the advantages of computer-aided testing may be apparent, the success of any testing program lies ultimately with its acceptance by patients and health care professionals. Examining client acceptance, Skinner et al. (1985) found that the acceptability of computerized testing increased after direct experience with a computer-administered test.[11] Staff acceptance of computers, however, poses a different problem. Staff members may consider automation a negative event, a threat to status, or even a threat to job security.[12] There could even be conflicts over how computers should be used in health care.[13] These often unstated conflicts need to be addressed with what Johnson and colleagues (1976) describe as "intentionally relevant systems."[14] Computerized applications developed in this context are designed to be based on organizational and individual needs. They are directly relevant to the kinds of decisions being made and are consistent with the operating style of the people using the system.

While computer testing in general has many advantages, these advantages may not hold for all assessment topics nor necessarily with all subject populations upon which the test may be applied. This chapter now turns to studies specifically addressing the use of computers for alcohol assessment. Two major questions are addressed: What effect does computerization have on the reliability and validity of alcohol assessment; and can computers be used in innovative ways to increase the consistency and accuracy of alcohol assessment?

Effect of Computers
on Alcohol Assessment

The first reported application of computer technology specifically to alcohol assessment was by Lucas et al. in 1977.[15] These researchers developed a computer-administered interview designed to assess an individual's history of alcohol use. Their results suggested that computer testing may elicit more accurate information than psychiatrist interviews. This study stimulated other research examining the same basic issues. These studies, summarized in Table 1, are grouped according to the target population that was employed: alcohol patients, psychiatric patients, family practice patients, and general community surveys.

Alcohol Patients

Lucas et al. (1977) developed a computerized test aimed at eliciting clinical information about alcohol use. Information included current alcohol consumption, history of alcohol use, and current symptoms. Thirty-six male volunteers, referred by general practitioners, agreed to participate in a special assessment procedure that involved one computer assessment and separate face-to-face interviews by two psychiatrists.

Results showed no difference between any of the three interviews in terms of identification of specific, alcohol-related symptoms. However, the computer elicited approximately 30% higher amounts of reported alcohol consumption than either psychiatrist interview. Assessment methods were also compared with the patients' concept of an ideal interview. In this comparison, one half of the patients preferred the computer interview, and an equal number rated the psychiatrist interview as closer to an ideal medical interview. Lucas et al. (1977) concluded that there was a high level of patient acceptance with computer testing. As well, there were some notable differences in reported alcohol consumption, with the computer reporting higher and likely, more accurate figures.[16]

Skinner and Allen (1983) examined alcohol, drug, and tobacco use in a randomized study that compared a computer assessment with face-to-face interview and self-report questionnaires.[16] Subjects were

Table 1
Comparison of Computerized, Interview, and Questionnaire Assessment

Reference/ study	Methods	Content	Finding
Alcohol patients			
15	Computer vs interview	Alcohol history	Computer ↑
16	Computer vs questionnaire vs interview	MAST (history or substance use)	No difference
17	Computer vs questionnaire	MMPI	No difference
18	Computer vs questionnaire	MAST (drinking habits and symptoms)	Computer ↑
Psychiatric patients			
19	Computer vs interview	Alcohol history	Computer ↑
Family practice patients			
11	Computer vs questionnaire	Various substance use measures	No difference
20	Computer vs interview	Various lifestyle topics including substance use	No difference
21	Computer vs questionnaire	Quantity and frequency of substance use	Some differences
Community survey			
22	Computer vs questionnaire	Frequency of alcohol and drug use	Computer ↓
23	Computer vs interview	Alcohol consumption	Computer ↑

randomly assigned to one of the three assessment formats ($n = 50$ per condition) and included male and female cases who had sought help at an alcohol- and drug-abuse treatment center. The assessment covered information about symptoms related to alcohol use, alcohol and drug consumption patterns, and perceptions of the assessment format.

Skinner and Allen (1983) found that the three different formats produced no differences in the reported level of alcohol problems nor in the client's consistency in answering items. In data examining substance use, no differences were reported across the three assessment formats, except in caffeine use, where small differences were reported in cups of tea drank and candy bars consumed. However, some differences were noted in the rating of different assessment methods. Specifically, computer testing was seen as more relaxing, fast, interesting, light, and short compared to other formats. The face-to-face interview was perceived as friendlier than the other formats. Of special note, no differences were perceived in terms of bad–good, pleasant–unpleasant, valuable–worthless, and inaccurate–accurate. Overall, this study suggested that the computer obtains the same accuracy when compared to other assessment techniques.

Another study conducted by Lambert et al. (1987) found no differences in computer vs questionnaire testing.[17] This study employed 68 alcoholics and seven drug abusers recruited of inpatient alcohol- or drug-treatment programs of a Veterans Administration medical center. Subjects completed the MMPI in two formats: computer version and paper-and-pencil questionnaire. Analysis examining the 14 MMPI scales found that the two methods were equivalent. The only significant difference between the two administration methods occurred in one scale—the "cannot say" scale; and this finding was attributed to eliminating the cannot say option from the computer version.

The final study using alcohol patients was conducted by Malcomb et al. (1989).[18] This study also used patients from a Veterans Administration medical center, selecting 50 patients admitted for alcohol treatment. Using the MAST as their primary measure, the researchers gave patients a computerized version and a questionnaire version. The MAST administrations occurred on two separate times: after a 5- to 7-d period of evaluation and alcohol/drug withdrawal and after a further 14 d of drug or alcohol rehabilitation treatment. Results showed higher MAST scores on the computerized version, with a mean score of 17.4 (SD = 4.5) vs the questionnaire's mean score of 16.2 (SD = 4.3). The Pearson correlation between the two versions was a modest 0.48 ($p < 0.01$).

Psychiatric Patients

Bernadt et al. (1989) examined the drinking history of 102 patients admitted to general psychiatric wards.[19] The assessment methods included: computer testing, face-to-face interviews with a nurse, and face-to-face interviews with a psychiatrist. The assessment examined the volume of alcohol consumed and responses to brief CAGE and MAST tests. The computer-elicited levels of consumption were similar to the nurse's ratings of alcohol consumption, and though the results were not identical, the authors pointed out that the level of computer agreement was as good as the agreement between nurses and psychiatrists. The authors also found that agreement decreased as the amount of alcohol intake reported increased, with nurses and psychiatrists reporting higher levels of consumption. In terms of MAST and CAGE items, no differences were found among the three assessment methods. Good agreement was indicated by a kappa of 0.77.

Family Practice Patients

Three studies have examined the effect of using computerized testing with patients from a family practice clinic. In the first study, Skinner et al. (1985) assessed 180 patients regarding their use of alcohol, tobacco, caffeine, medications, and nonmedical drugs.[11] This assessment was completed by randomly assigning subjects to computerized testing, paper-and-pencil questionnaires, or face-to-face interviews. Patients rated their preference for all three methods before and after testing. Analysis indicated that subjects preferred face-to-face interviews over the other methods. However, after actual testing, there was a threefold increase in the preference for computerized testing. In terms of reported substance use, there were no significant differences between the three methods.

The second study examined patients' consistency in reporting information to the computer and to their doctor.[20] Using 117 family practice patients, subjects were assessed using two methods: computerized testing and a questionnaire administered in a face-to-face meeting by the patient's family physician. In addition, the computerized test was administered on separate occasions approximately 3 wk apart in order to assess test–retest reliability. The assessment covered a variety of lifestyle issues including alcohol, tobacco, and caffeine use, weight, sleep habits, physical activity, and seatbelt use. These researchers found good retest reliability for computerized testing. The two separate occasions had a median test-retest correlation of 0.85 for 20 lifestyle factors. In addition, there was good consistency between the computer-evoked lifestyle concerns and those reported to the physician. The median proportion of agreement was 0.92 with a median kappa statistic of 0.75.

Bungey and coworkers (1989) replicated aspects of Skinner et al. (1985) by examining the effects of different assessment methods using 300 patients attending a family practice unit in Australia.[11,21] This study used three methods of testing: computerized testing, paper-and-pencil questionnaires, and face-to-face interviews. The assessment covered problem recognition, degree of substance consumption, and consequences of substance use for each category of substance (alcohol, tobacco, and other drugs). There were some differences reported when comparing the effect of different assessment methods on reported alcohol use. However, the authors suggested that there was no meaningful pattern to the results. One finding suggested higher tobacco problem recognition with computerized testing. In terms of the patients' preferred method of assessment, face-to-face interviews were considered best; and patients' acceptance increased after exposure to computerized testing. These findings largely replicated those of Skinner et al. (1985).[11]

Community Surveys

Erdman et al. (1983) questioned 2300 students about alcohol and drug consumption.[22] Students were randomly assigned to one of two methods: a computerized test or a paper-and-pencil test. A subsample of 133 subjects was given both versions in random order. Results indicated high agreement between the two methods (with kappa ranging from 0.65 to 0.86). Alcohol use, however, showed somewhat less agreement than tobacco and drug use, with much of the disagreement the result of direct contradictions in the two forms (saying "yes" to alcohol use in one method and "no" in the other). However, the authors pointed out that similar inconsistencies occur when repeating paper-and-pencil questionnaires. Alcohol showed less consistency than tobacco and drug use when considering frequency of use. In total, the authors concluded that the two methods produced very similar results, with the stipulation that the computer interview elicited lower alcohol consumption and marijuana use.

A survey of subjects in their home was completed by Duffy and Waterton (1984).[23] These researchers compared computer assessment vs face-to-face interviews in an examination of 320 males in Edinburgh, UK. This investigation focused on the amount of alcohol consumed in the previous 7 d. Analysis suggested that computer assessment resulted in higher reports concerning the amounts of alcohol consumed. Overall, this amount was 33% higher than face-to-face questioning. The study also found that computer testing was greeted with a high degree of acceptance concerning issues such as concentration required, accuracy, and ease of test completion.

Overall, the studies in Table 1 indicate an absence of consistent difference when computerized testing is compared to personal interviews and pencil-and-paper questionnaires. However, these studies have only used the computer to mimic an interview or questionnaire and have not capitalized upon the rapid decision-making power of the technology. The next section explores some of these unique advantages.

Response Latency:
An Innovative Use of Computers

A unique advantage of computerized assessment is its ability to measure otherwise inaccessible information. In particular, the computer can measure how long the individual waits before answering a question, response latency. By analogy, one might think of an interviewer being aware that a question was answered either too quickly or too slowly and using this information to probe further the interviewee. The computer, however, can accurately measure the exact length of the delay.

Theoretically, Erdle and Lalonde (1986) describe the usefulness of latency measures in terms of cognitive prototypes.[24] An especially important prototype concerns information about the self. This prototype is used in all self-relevant decisions, particularly those in which individuals are asked to state whether a trait characteristic describes them. Erdle and Lalonde argue that people high in a given trait, e.g.,

Gavin, Skinner, and George

STIMULUS

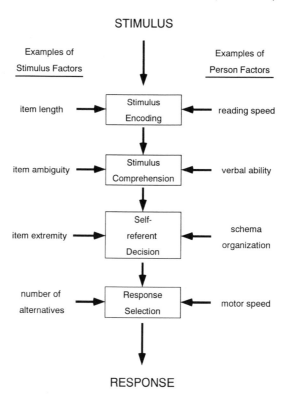

Fig. 1. Person and stimulus factors affecting response latency.[25]

severe degree of alcohol dependence, will show different reaction
latency patterns compared to those low in a trait, e.g., mild level of
alcohol dependence. High-trait individuals will respond faster to items
they agree with, such as major alcohol withdrawal symptoms, and
slower to items they do not endorse.

Holden et al. (1991) emphasize, however, that response latency
is affected at several points in the response process (Fig. 1).[25]
Describing the cognitive process involved in answering a questionnaire item, they emphasize four steps: initial stimulus encoding of the
item; comprehension of its content; the self-referent decision; and
final response selection. Each step is affected by factors related to the
stimulus (the item itself) and the person responding. For example,
encoding is affected by item length and the person's reading speed.
Comprehension is affected by item ambiguity and the subject's

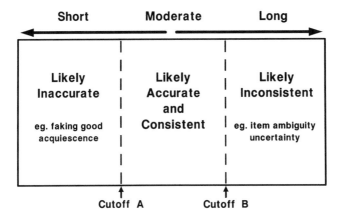

Fig. 2. Model of response latency for distinguishing accurate from inaccurate responses.

reading speed. The self-referent decision is affected by item extremity and the person's cognitive prototype. Finally, response selection can be affected by the number of alternatives and the subject's motor speed.

The latency we want to know the most about concerns the self-referent decision: That is, how long does the person take to decide if the item is self-descriptive? The rest of the factors affecting reaction latency are confounds, only adding errors that prevent us from accurately measuring latency of self-referent decisions. For example, in clinical patients undergoing withdrawal from alcohol or other drugs, reading speed could be decreased or inconsistent, motor responses could be erratic, and attention to the test process could wander.

According to findings from a variety of studies, response latency has been shown to be an important indicator of inconsistency and inaccuracy of item responses.[24–28] Figure 2, adapted from George (1990), summarizes a model describing the relationship of item accuracy and consistency with response latency.[10] In this model, short latencies are associated with inaccurate measures, whereas long latencies are associated with inconsistent responses.

Research suggests that individuals responding to items with the goal of appearing in a more favorable, socially desirable manner ("faking good") have shorter reaction time. For example, Fig. 3 presents results from George and Skinner (1990) in which subjects were instructed to respond honestly to a computerized lifestyle assessment

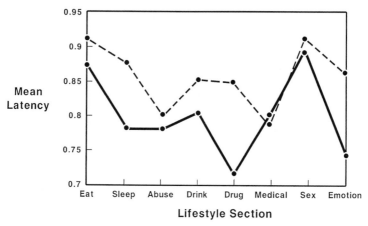

Fig. 3. Mean latency scores for "fake-good" vs "honest" assessment conditions.

vs instructions to "fake good" and present a favorable image.[29] Seventy clients from an addiction treatment center were asked to complete the CLA under two instructional conditions: "fake good" and "honest" self-report. An equal number of participants were randomly assigned to each group in a counterbalanced design. Results indicated that the combined group latency scores for the "fake good" respondents were faster than those for honest self-report; this is partly an artifact of the instruction order. Subjects were generally faster during their second self-report regardless of the type of instruction given. However, in the area of emotional health, subjects were faster for faking good regardless of instruction order.

Increasing the Accuracy of Assessment

Response latency may also prove valuable for increasing the accuracy of assessment. This increase can be gained by flagging items in which the response latency is either too short or too long. By examining the distribution of latency scores to a specific item across all subjects, a latency cutoff time can be set. Beyond the cutoff boundaries, any response that is too quick or too slow can be considered suspect (Fig. 2).

Empirical support for this model is provided by George (1990).[10] Subjects were classified by their instruction condition, either fake good or honest self-report. Different cutoff times were then evaluated in

Table 2
Decision Analysis of Latency Cutoffs*

	Assessment instruction	
Prediction from latency scores	"Fake good,"(†) positive	"Honest," negative
Positive, below cutoff	True positive (A)	False positive (B)
Negative, above cutoff	False negative (C)	True negative (D)

*Definition of criteria: sensitivity = A/(A + C); specificity = D/(B + D); positive predictive value = A/(A + B); negative predictive value = D/(D + C); and overall accuracy = A + D/(A + B + C + D).

terms of how well latency scores differentiate responses in the honest group compared to those under the fake good condition. Clinical decision analysis was used to assess the efficiency of different cutoffs (in essence, moving cutoff point A further to the right side in Fig. 2).[30] At each cutoff, subjects were classified by their known test-response conditions (fake good or honest response). Item responses were also classified as deviant or accurate using the response latency cutoff score. These assignments result in a two-by-two cross-classification depicted in Table 2.

An ideal latency cutoff would correctly separate the honest responses from the fake responses: high levels of true positive and true negative decisions, and low levels of false decisions. Table 3 summarizes the results of using different cutoff scores. At strict cutoffs (5th percentile), 28% of the fake good responses are identified, and only 10% of the honest responses are misidentified. The overall accuracy reaches 59% of the decisions. As latency cutoff is increased to the 35th percentile, overall accuracy peeks at approximately 64%.

In addition to the fake good response patterns studied in the above section, short reaction times also may be indicative of acquiescence responses (tendency to agree with items), deviant response patterns (providing extreme answers that ignore item content), denial (minimizing problems), and carelessness. All of these response patterns are threats to the accuracy of a test and can be identified at least partially by latency measurement. These unusual responses could be used to cue further items for probing or indicate sections of the assessment

Table 3
Accuracy of Latency Cutoff Scores in Detecting "Fake Good" Response

Latency cutoff point, percentile	Criteria				
	Sensitivity %	Specificity %	Positive predictive value %	Negative predictive value %	Overall accuracy %
5th	28	90	75	56	59
10	35	82	66	56	59
15	45	75	65	58	60
20	53	70	64	60	61
25	57	66	63	60	61
30	62	63	63	63	63
35	67	61	63	65	64
40	74	54	62	68	64

that should be discounted. In this way, response latencies and subsequent probes can be used in real time as the assessment proceeds in order to increase accuracy.

Conclusions

This review indicates that computerized testing is generally as accurate as more traditional interview and questionnaire approaches. Moreover, there is suggestive evidence from some studies that computers may elicit more accurate assessment of the quantity and frequency of alcohol consumption. However, the findings are not conclusive: Some studies have found that the methods elicit similar alcohol consumption levels, but others suggest different levels (Table 1). Further research is needed to understand the conditions under which a particular assessment approach, e.g., computer, may be expected to yield more reliable and valid information. Factors to be investigated include the assessment context (e.g., clinical vs community setting), reasons for the assessment (e.g., voluntarily seeking help vs mandatory, court-ordered assessment), physical and mental status of the respondent, safeguards regarding confidentiality of information, demographic characteristics (e.g., age), and previous experience with computers.

In addition, computerized assessment offers benefits that go beyond traditional methods. These include the ability to provide instant feedback throughout the assessment process and to generate a detailed report incorporating "expert advice" upon completion of the assessment. The decision-making power of the computer is a unique benefit that we are only beginning to exploit. To our knowledge, no study has been conducted to date in the alcohol treatment field that has evaluated the sophisticated statistical models and methodologies developed for adaptive or tailored testing.[31,32] Another exciting innovation described in this chapter is the use of response latency measures to indicate when a respondent may be giving inaccurate or inconsistent data (Fig. 2). Adaptive testing and the use of response latency represent significant opportunities for increasing the accuracy of assessment of alcohol use.

References

[1]M. S. Moncher, C. A. Parms, M. A. Orlando, S. P. Schinke, S. O. Miller, J. Palleja, and M. B. Schinke (1989) Microcomputer-based approaches for preventing drug and alcohol abuse among adolescents from ethnic-racial minority backgrounds. *Comput. Human Behav.* **5,** 79–93.

[2]W. V. Slack, G. P. Hicks, C. E. Reed, and L. J. Van Cura (1966) A computer-based medical-history system. *N. Engl. J. Med.* **274,** 194–198.

[3]J. N. Butcher (1987) The use of computers in psychological assessment: An overview of practices and issues, *Computerized Psychological Assessment: A Practitioner's Guide.* J. N. Butcher, ed. (Basic Books, New York).

[4]T. R. Kratochwill, E. J. Doll, and W. P. Dickson (1985) Microcomputers in behavioral assessment: Recent advances and remaining issues. *Comput. Human Behav.* **1,** 277–291.

[5]A. C. Carr, A. Ghosh, and R. J. Ancill (1983) Can the computer take a psychiatric history? *Psycholog. Med.* **13,** 151–158.

[6]W. M. Evan and J. R. Miller (1969) Differential effects on response bias of computer versus conventional administration of a social science questionnaire. *Behav. Sci.* **14,** 216–227.

[7]M. D. Reckase (1977) Procedures for computerized testing. *Behav. Res. Methods Instrum.* **9,** 148–152.

[8]J. P. Spinetti and E. K. Hambleton (1977) A computer simulation study of tailored testing strategies for objective-based instructional programs. *Educat. Psycholog. Meas.* **37,** 139–158.

[9]R. A. English, M. D. Reckase, and W. M. Patience (1977) Application of tailored testing to achievement measurement. *Behav. Res. Methods Instrum.* **9,** 158–161.

[10]M. S. George (1990) *The use of response latency to study accuracy and consistency in a computerized lifestyle assessment.* Doctoral dissertation (University of Toronto, Toronto, Canada), unpublished manuscript.

[11]H. A. Skinner, B. A. Allen, M. C. McIntosh, and W. H. Palmer (1985) Lifestyle assessment: Just asking makes a difference. *Brit. Med. J.* **290,** 214–216.

[12]B. Kleinmuntz (1975) The computer as clinician. *Am. Psycholog.* **30,** 379–387.

[13]H. P. Erdman, J. H. Greist, M. H. Klein, J. W. Jefferson, and C. Getto (1981) The computer psychiatrist: How far have we come? Where are we heading? How far dare we go? *Behav. Res. Methods Instrumen.* **13,** 393–398.

[14]J. H. Johnson, R. A. Guinnetti, and T. A. Williams (1976) Computers in mental health care delivery: A review of the evaluation toward interventionally relevant on-line processing. *Behav. Res. Methods Instrumen.* **8,** 83–91.

[15]R. W. Lucas, P. J. Mullin, C. B. Luna, and D. C. McInroy (1977) Psychiatrists and a computer as interrogators of patients with alcohol-related illnesses: A comparison. *Brit. J. Psychiat.* **131,** 160–167.

[16]H. A. Skinner and B. A. Allen (1983) Does the computer make a difference? Computerized versus face-to-face versus self-report assessment of alcohol, drug and tobacco use. *J. Consult. Clin. Psychol.* **51,** 267–275.

[17]M. E. Lambert, R. H. Andrews, K. Rylee, and J. R. Skinner (1987) Equivalence of computerized and traditional MMPI administration with substance abusers. *Comput. in Human Behav.* **3,** 139–143.

[18]R. Malcomb, E. T. Sturgis, R. F. Anton, and L. Williams (1989) Computer-assisted diagnosis of alcoholism. *Comput. Human Serv.* **5,** 163–170.

[19]Bernadt, Daniels, Blizard, and Murray (1989) Can a computer reliably elicit an alcohol history? *Brit. J. Addict.* **84,** 405–411.

[20]H. A. Skinner, W. Palmer, M. Sanchez-Craig, and M. McIntosh (1987) Reliability of a lifestyle assessment using microcomputers. *Can. J. Public Health* **78,** 329–334.

[21]J. B. Bungey, R. G. Pols, K. P. Mortimer, O. R. Frank, and H. A. Skinner (1989) Screening alcohol and drug use in a general practice unit: Comparison of computerized and traditional methods. *Comm. Health Stud.* **13,** 471–483.

[22]H. P. Erdman, M. H. Klein, and J. H. Greist (1983) The reliability of a computer interview for drug use/abuse information. *Behav. Res. Methods Instrumen.* **15,** 66–68.

[23]J. C. Duffy and J. Waterton (1984) Under-reporting of alcohol consumption in sample surveys: The effect of computer interviewing in fieldwork. *Brit. J. Addict.* **79,** 303–308.

[24]S. Erdle and R. N. LaLonde (1986) *Processing information about the self: Evidence for personality traits as cognitive prototypes.* Paper presented at the Canadian Psychological Association Annual Convention, Toronto.

[25]R. R. Holden, G. C. Fekken, and D. H. G. Cotton (1991) Assessing psychopathology using structured test-item response latencies. *Psycholog. Assess.: J.Consult. Clin. Psychol.* **3,** 111–118.

[26] G. C. Fekken and R. R. Holden (1988) *Response latency evidence for viewing personality dimensions as schema.* Paper presented at the 96th Annual Convention of the American Psychological Association, Atlanta, GA.

[27] R. R. Holden and G. C. Fekken (1987) *Reaction time and self-report psychopatholoaical assessment: Convergent and discriminant validity.* Paper presented at the 95th Annual Convention of the American Psychological Association, New York.

[28] S. M. Popham and R. R. Holden (1990) Assessing MMPI constructs through the measurement of response latencies. *J. Person. Assess.* **54**, 469–478.

[29] M. S. George and H. A. Skinner (1990) Using response latency to detect inaccurate responses in a computerized lifestyle assessment. *Comput. Human Behav.* **6**, 167–175.

[30] M. C. Weinstein, H. V. Fineberg, A. Elstein, H. Frazier, D. Neuhauser, R. Neutra, and B. McNeil (1980) *Clin. Decision Anal.* (W. B. Saunders, Toronto).

[31] D. J. Weiss and C. D. Vale (1987) Computerized adaptive testing for measuring abilities and other psychological variables, *Computerized psychological assessment: A practitioner's guide.* J. N. Butcher, ed. (Basic Books, New York).

[32] D. J. Weiss (1985) Adaptive testing by Computer. *J. Consult. Clin. Psychol.* **53**, 774–789.

Timeline Follow-Back

A Technique
for Assessing Self-Reported Alcohol Consumption

Linda C. Sobell and Mark B. Sobell

Introduction

Although the measurement of drinking is necessary for assessing and evaluating the treatment of alcohol problems, this key dependent variable has not always been reported in outcome studies.[1-3] Today, the issue is not whether to measure drinking, but how to measure drinking. Concerns about how best to measure drinking patterns and problems date back to at least 1926, when Pearl stressed the importance of separating steady daily drinkers from occasional heavy drinkers.[4]

Measuring drinking is not simple, and there is no single, preferred procedure. In a recent review, Room (1990) assesses the development of and rationales for various methods that have been used to measure alcohol consumption.[5] He classifies the methods as employing one of two major strategies: (a) having respondents report all recent drinking days (recent occasions) or (b) having respondents summarize their current drinking pattern. This latter method is often referred to as an aggregate measure of consumption.

In North America, summary or aggregate measures have long predominated. Several variations of these measures have been developed over the years. Typically labeled quantity–frequency (QF) methods, they have been soundly and repeatedly criticized as being

From: *Measuring Alcohol Consumption*
Eds.: R. Litten and J. Allen ©1992 The Humana Press Inc.

"relatively insensitive to differences in the patterning of drinking that are of social importance in terms both of correlates and of consequences... More crucially, many of the adverse consequences of drinking—notable casualties and social problems—are fairly specific to heavy drinking occasions".[5(p67)–11]

Although the appeal of using QF estimation formulae is easily understood—they provide quick, easy measurement of consumption—their insensitivity to atypical heavy drinking and to variable patterns of drinking has forced researchers to look for alternative, more precise methods. For example, one alternative has been to add questions to existing QF instruments. However, this lengthens administration time and therefore obviates one of the major advantages of QF methods.[10] Whatever the alternative, when drinking has been more precisely assessed, considerably more ethanol consumption has been reported.[5,11] Compared to simple QF assessments, studies that have added questions assessing heavy drinking with normal drinkers have reported an increase in the overall average volume of reported drinking ranging from 16 to 36%.[5] Similar findings have been observed for problem drinkers.[10,12,13] Further, a recent study demonstrated that beverage-specific QF questions (e.g., different questions for beer, wine, and hard liquor) yielded much higher estimates (47%) of ethanol consumption than global QF questions, especially among heavy drinkers and individuals who consumed more than one beverage.[14] The major conclusion of these studies is that atypical, combined beverage use and non-patterned drinking cannot be ignored.

Development of the TLFB Method

More than two decades ago, the timeline was developed as a procedure to aid recall of past drinking. That method, first referred to as the gathering of daily drinking disposition data and later labeled as the timeline follow-back (TLFB) method, is the focus of this chapter.[15] Currently, self-reports are the only viable method for retrospectively measuring drinking with any precision. Alternative methods exist, but they are either impractical (e.g., continued direct observation), fraught with problems (e.g., alcohol sweat patches) or they only measure very recent drinking (e.g., biochemical indicators or blood alcohol tests.[16–20] Consequently, there is no practical alternative technology other than self-reports for retrospectively assessing drinking. TLFB stands as the most exhaustively evaluated method that can be used for this purpose.

Development of the TLFB method was prompted by the investigation of alternative treatment goals to abstinence. In the early 1970s, there was not only a paucity of measures to assess drinking, but reports of drinking were often classified using rather crude categories, e.g., improved, same, or worse than prior to treatment.[2,21–23] This measurement approach was largely a result of conventional wisdom of the day, which held that it was only important to assess whether or not the person was drinking, with the corollary being that any drinking was clear evidence of failure.[24] Such imprecise measures presented two problems for evaluating moderation outcomes.[2,21,22] First, unlike abstinent outcomes, for which an all-or-none judgment is sufficient, moderation outcomes require a specification of the amount consumed. Second, since it was clear that moderation outcomes would be scrutinized, their measurement had to be very precise.[25,26] Although it had not been done before, it appeared that the most straightforward way to assess drinking was to ask subjects to recall as accurately as possible the number of days they had consumed alcohol and the amount they had consumed on each of those days. TLFB, therefore, is a direct, parsimonious approach toward retrospectively measuring drinking. Its embellishment has been mainly in terms of procedures to aid recall.

Administration of the TLFB Technique

TLFB has been evaluated and discussed in several major reports in the literature.[10,11,27–34] In addition, a computerized version of TLFB with a detailed manual is in preparation.[35] In this section, key features and elements of the technique will be described.

The TLFB method presents subjects with a calendar and asks them to provide retrospective estimates of their daily drinking over a specified time period ranging up to 12 mo. Prior to administration of TLFB, subjects should be breath tested to insure that they are alcohol-free.[17,36] For clinical populations, it takes about 25–30 min to gather 12 mo of data and considerably less time for shorter intervals (e.g., approximately 10 min for 90 d). TLFB can be either administered by

Table 1
TLFB Interview Techniques

Daily Calendar is a major aid that assists recall of drinking.

Key Dates uses holidays, birthdays, newsworthy events, and idiosyncratic happenings to assist recall of drinking.

Standard Drink Conversion is an aid to report combined beverage use and/or nonstandard consumption amounts.

Black and White Days identify periods of invariant drinking and extended abstinence.

Discrete Events is the use of specific events to identify periods of drinking or nondrinking (e.g., arrests, hospitalizations, illnesses, employment, or treatment participation).

Anchor Points identify drinking behavior that anchors each event or major drinking episode.

Boundary Procedure establishes upper and lower limits for reporting amount consumed (i.e., when starting the interview, first ask about the most or least amount consumed on a single day during the reporting period).

Exaggeration Technique is used to avoid vague drinking labels or descriptions (e.g., a lot or a little drinking). Present an exaggerated minimum or maximum value to help specifically define the label.

---

an interviewer or self-administered. Several memory aid procedures have been developed to help people recall their drinking when completing the TLFB. Briefly, these aids, described in Table 1, include:

1. Using a visual calendar,
2. Listing key dates on the calendar,
3. Using a standard drink-conversion card,
4. Identifying periods of invariant drinking or extended abstinence,
5. Using key dates and periods of invariant drinking to provide anchors for reporting drinking that occurred around these episodes,
6. Using a boundary procedure, and
7. Using an exaggeration technique.

Interview Date: June 9/90   Date of Last Drink: June 8/90
Follow-back (60 days): April 9/90 thru June 7/90

1990

APRIL

	SUN	MON	TUE	WED	THU	FRI	SAT
A	1	2	3	4	5	6	7
P	8	9 5	10 6	11 8	12 6	13 12	14 16
R	15 6	16 8	17 6	18 10	19 0	20 8	21 12
I	22 15	23 10	24 12	25 9	26 8	27 15	28 17
L	29 8	30 6					

MAY

	SUN	MON	TUE	WED	THU	FRI	SAT
			1 8	2 8	3 10	4 12	5 15
M	6 7	7 8	8 0	9 6	10 9	11 15	12 17
A	13 12	14 6	15 8	16 15	17 8	18 9	19 10
Y	20 8	21 8	22 9	23 10	24 15	25 8	26 12
	27 5	28 6	29 12	30 8	31 9		

JUNE

	SUN	MON	TUE	WED	THU	FRI	SAT
						1 10	2 15
J	3 0	4 10	5 6	6 0	7 12	8 Last Drink	9 Today
U	10	11	12	13	14	15	16
N	17	18	19	20	21	22	23
E	24	25	26	27	28	29	30

Fig. 1. Sample timeline follow-back calendar.

A very recent study provides strong support for using aids to assist the recall of drinking.[37] Subjects were asked to think aloud about their responses to four alcohol questions; i.e., average quantity, frequency of drinking, frequency of drinking over five drinks, and frequency of drunkenness. The results indicate that subjects most often used "anchoring and adjustment" and "context" strategies to arrive at their answers. It is interesting that the authors suggest that these findings can help researchers "improve accuracy of responses by providing cues to assist in recall,"[37(p251)] as the TLFB method uses techniques to aid recall of drinking. Sample instructions for administering TLFB appear in Appendix A. These instructions can be modified for different target groups or research projects.

The two most significant memory aids are the daily calendar and the use of a standard drink format. A sample calendar is shown in Fig. 1. Key dates such as national holidays and significant personal events (e.g., anniversaries, birthdays of self and family members, or vacations) as well as newsworthy events (e.g., presidential election, or Gulf war) can be noted on the calendar. The calendar provides subjects with a temporal framework for recalling dates and patterns

1 Standard Drink (13.6 g absolute alcohol) =

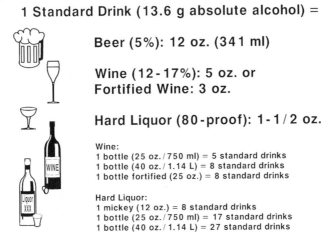

Beer (5%): 12 oz. (341 ml)

Wine (12-17%): 5 oz. or
Fortified Wine: 3 oz.

Hard Liquor (80-proof): 1-1/2 oz.

Wine:
1 bottle (25 oz./750 ml) = 5 standard drinks
1 bottle (40 oz./1.14 L) = 8 standard drinks
1 bottle fortified (25 oz.) = 8 standard drinks

Hard Liquor:
1 mickey (12 oz.) = 8 standard drinks
1 bottle (25 oz./750 ml) = 17 standard drinks
1 bottle (40 oz./1.14 L) = 27 standard drinks

Fig. 2. Sample standard drink-conversion card.

related to their drinking, e.g., bowling on Thursdays or playing cards on Wednesdays. Also, the time period over which drinking is to be recalled is marked on the calendar and reviewed with the subject. Some subjects have even requested and found it helpful to use their own appointment books or personal date books or calendars as aids in recalling their drinking. Use of any of these or any other materials that assist recall is encouraged.

The second significant aid involves having subjects report their drinking in terms of standard drinks. A sample copy of a standard drink-conversion card is shown in Fig. 2. Investigators needing different drink-conversions (e.g., 100 proof spirits) can design a standard drink-conversion card to meet their needs. Subjects are quickly taught the drink equivalencies, and they have had little difficulty using this method to report their drinking. This procedure also facilitates reporting the use of more than one type of beverage on each drinking occasion (e.g., two beers and one glass of wine), as occurs in a sizable percentage of different drinker groups.[10,11,14,34]

Features

TLFB has several valuable features: The time period over which data can be collected ranges up to 12 mo. Although going beyond this limit might be feasible, the technique has not been empirically evaluated for longer intervals. The TLFB method also allows several dimensions of a subject's drinking to be separately examined: variability (i.e., scatter); pattern (i.e., shape); and extent of drinking (i.e., elevation, or how much). In addition, TLFB can generate a variety of continuous variables that provide quite different and more precise information about an individual's drinking than estimation formulae. For example, variables that can be generated include but are not limited to the following:

1. Percent of days drinking at various levels—e.g., abstinent days, high-consumption days, low-consumption days,[10,11,34]
2. Mean number of drinks each drinking day,
3. Maximum number of drinks consumed on any single day, and
4. Temporal patterns of drinking—weekend/weekday, binges.[34]

Data from TLFB can also be used to identify periods of relapse and can serve as the basis for probing antecedents of relapse.[38] Aside from relapse, by identifying atypical heavy drinking days, TLFB allows one to relate such drinking to health risks and other serious consequences.[10] As is evident, a key feature of TLFB is that it provides a detailed, clinically useful picture of the full range of a subject's drinking.

Finally, a common reaction among those who have not previously used TLFB has been to consider its administration as very exacting or even impossible. However, our experience, as well as that of others who have used the procedure, has been that with a small amount of training, it is quite easy to learn. Although the task may seem intimidating at first, most concerns quickly dissipate with practice. The cliché *Try it, you'll like it* may be appropriate here. The fact that the technique has been used by many researchers and in several countries clearly supports its feasibility.

Psychometric Properties

The psychometric characteristics of any assessment instrument are very important because they determine the confidence one can have in data gathered by that instrument. In this regard, the TLFB method has been shown to have very good psychometric characteristics with a variety of drinker groups. Four aspects of the psychometric properties of TLFB will be discussed: test–retest reliability, comparison of subjects' (S) and collaterals' (C) reports of the subject's drinking, comparison of subjects' TLFB reports of incarcerations compared to official record data, and concurrent validity.

Test–Retest Reliability

It has been asserted that "if a test does not even correlate with itself when administered on two occasions, it is hopeless to employ the test in correlational studies."[39(p234)] Thus, a prerequisite to establishing an instrument's validity is the demonstration of its reliability. Table 2 presents TLFB test–retest reliability correlation coefficients from several studies. Over the course of 10 yr, the method has been evaluated across multiple drinker populations—high- and low-dependence problem drinkers, male and female normal drinkers in the general population, and black and white male and female college students. Overall, the table shows that drinking data derived from the TLFB method generally have high test-retest reliability across multiple populations of drinkers.

Subject–Collateral Comparisons

Table 3 summarizes results from six clinical studies that have compared S and C reports of a subject's drinking. The correlations presented are for pre- and posttreatment intervals ranging from 12 mo pretreatment to 18 mo posttreatment across five drinking variables. Whereas moderate correlations were observed for S and C reports of high and low drinking days, very high correlations occurred for total drinking days, jail days, and hospital days across all studies.

Validity of Subjects' Reports of Days Jailed, Hospitalized, and in Residential Treatment

Besides collecting reports of the amount of drinking per day, TLFB can also reflect days that subjects report having spent in jail, in a hospital, or in a residential treatment facility. Studies that have collected such data provide some evidence for the validity of the TLFB method. When inpatient and outpatient alcohol abusers' TLFB reports of days jailed, hospitalized, or at residential treatment facility pre- and posttreatment have been compared with official records (i.e., external validity criterion), the results generally indicate a high

Table 2
Test-Retest Reliability of TLFB Method Across Multiple-Subject Populations

Variable	Normal drinkers (62)	College students (80)	College students (48)	Residential alcoholics (12)	Outpatient alcoholics (12)	Inpatient alcoholics (12)	Veteran alcoholics (26)
				Population type (*n*)			
Days abstinent							
90	0.96	0.96	0.96	0.62	0.85	0.94	0.94
30	0.90	0.92	0.93	0.69	0.79	0.88	0.93
Days 1–6 drinks							
90	0.95	0.96	0.93	0.67	0.90	−0.13	0.83
30	0.88	0.92	0.87	—†	0.92	−0.13	0.85
Days >6 drinks							
90	0.94	0.95	0.97	0.43	0.91	0.60	0.91
30	0.86	0.96	0.97	0.62	0.88	0.50	0.92
Reference	11	34	72	73	73	73	74

*Reprinted by permission from *British Journal of Addiction*. Original table appeared in a previous article on the timeline method.[11]
†Correlation could not be computed because no residential subject reported any days of 1–6 drinks during the 30-d interval.

Table 3
TLFB Comparison of S and C Reports of Subject's Drinking

Variable	Reference (n = Subject–Collateral pairs)					
	[30]Maisto et al. 1979, n = 46*	[75]O'Farrell et al. 1984, n = 73	[40]Maist et al. 1985, n = 34	[76,77]McCrady et al. 1986,1991, n = 42	[78]Zweben 1986, n = 87	[79]Zweben et al. 1988, n = 116
Treatment interval (mo)	Post (6)	Pre (12)	Post (18)	Post (6)	Pre (12)	Post (18)
Abstinence						
Correlation (r)	0.81	0.88	0.86	0.92	0.79	0.92
Mean days (S/C)	106/105	133/130	385/366	75/61	121/121	—
Low alcohol consumption						
Correlation (r)	0.49	0.38	0.95	—	0.41	0.74
Mean days (S/C)	5/3	45/30	81/91	—	78/91	—
Amount†	1–6	1–6	1–6	1–2	1–4	1–4
High alcohol consumption						
Correlation (r)	0.82	0.59	0.68	—	0.65	0.67
Mean days (S/C)	55/64	159/153	54/62	17/22	161/149	—
Amount†	≥7	≥7	≥7	≥3	≥5	≥5
Jail (alcohol) consumption						
Correlation (r)	0.46	0.99	0.94	—	—	—
Mean days (S/C)	4/2	3/3	8/4	—	—	—
Hospital (alcohol)						
Correlation (r)	0.97	0.98	—	—	—	—
Mean days (S/C)	10/7	27/28	—	—	—	—

*For correlations, n = 46; for mean days, n = 46 for S, and n = 51 for C.
†Amount in number of standard drinks.

degree of correspondence between subjects' self-reports and official records.[27,28,40] This conclusion holds for both duration (i.e., number of days confined on each occurrence) and incidence (i.e., number of confinements) data. In almost all cases, when discrepancies occurred, it was because subjects overreported days confined as compared to official records.

Concurrent Validity

The third way the psychometric properties of TLFB were evaluated was by comparing TLFB data with two established measures of alcohol-related disabilities—Alcohol Dependence Scale (ADS), Short Michigan Alcohol Screening Test (SMAST)—and two biochemical tests frequently used to assess alcohol-related acute hepatic dysfunction—γ-gamma-glutamyl-transpeptidase (γ-GGT) and serum glutamic-oxaloacetic-transaminase (SGOT).[41–46] Correlations between drinking variables derived from TLFB and scores on the ADS and SMAST are shown in Table 4. Table 5 presents correlations between TLFB and the biochemical tests.

As shown in Table 4, higher ADS and SMAST scores were positively and significantly correlated with several TLFB drinking variables: a higher level of overall consumption, a higher number of heavy drinking days, and a greater number of mean drinks each drinking day. Also, low-consumption days were significantly but negatively correlated with the ADS and SMAST. These correlations suggest that the level of alcohol problems or dependence is directly related to reports of frequent heavy drinking as assessed by TLFB.

The liver function data in Table 5 are from a study of alcohol abusers seen in an outpatient treatment program (Nashville, TN). A standard part of the intake assessment included obtaining a comprehensive drinking history and the conduct of a brief medical evaluation that included a blood test. Subjects were selected for having γ-GGT values either within the normal range ($n = 22$; 0–50 mu/mL) or three or more SDs beyond the normal range ($n = 15$; > 50 mu/mL). As shown in Table 5, all but one of the TLFB drinking variables for both the 30 and 60 d prior to the blood test were significantly ($p < 0.05$), albeit moderately, and positively correlated with blood test values. These findings suggest that acute hepatic functioning is systematically related to heavy frequent drinking as reported on the TLFB.

Table 4
TLFB Data Correlated with Two Scales of Alcohol-Related Disabilities

Variable	Study 1(n = 56)[*] ADS[‡]	Study 2 (n = 118)[†] ADS[‡]	Study 2 (n = 118)[†] SMAST[§]
Heavy-consumption days[‖]	0.53[**]	0.62[**]	0.51[**]
Low consumption days[¶]	−0.38[**]	−0.26[**]	−0.39[**]
Total drinking days[#]	−0.14	0.38	0.25[**]
Mean drinks/drinking day	0.52[**]	—	—
Total drinks	0.37[**]	—	—

[*]Study 1, Ref. 35. [†]Study 2, Ref. 34. [‡]ADS = Alcohol Dependence Scale (Ref 41). [§]SMAST = Short Michigan Alcohol Screening Test (Ref 42). [‖]Study 1: ≥ 10 standard drinks (SD); Study 2: > 12 SD. [¶]Study 1: 1–4 SD; Study 2: 1–3 SD. [#]Study 1: 360 d pretreatment; Study 2: 30 d preinterview. [**]p < 0.01

Table 5
TLFB Data
Correlated with Two Biochemical Tests with 37 Alcohol Abusers[*]

Variable[*]	Serum glutamic- oxaloacetic transaminase, SGOT	Serum gamma- glutamyl transpeptidase, γ-GGT
Frequency of days drinking, 30 d	0.42[†]	0.36[†]
Frequency of days drinking, 60 d	0.41[†]	0.36[†]
Amount consumed, 30 d	0.34[†]	0.36[†]
Amount consumed, 60 d	0.39[†]	0.33[†]
Drinks/drinking day, 30 d	0.34[†]	0.35[†]
Drinks/drinking day, 60 d	0.36[†]	0.24

[*]Drinking data were calculated back from the date of the subject's blood test. [†]p < 0.05.

When the two groups, normal GGT (NGGT) and abnormal GGT (AGGT), were compared on these same variables using t-tests (two-tailed) for independent samples, the only significant differences occurred for frequency of drinking [Mean (SD) days drinking during last 60 d: NGGT = 28.0 (21.1) and AGGT = 41.1 (14.4); p = 0.04. Mean (SD) days drinking during last 30 d: NGGT = 13.0 (11.3), and AGGT = 20.8 (7.2); p = 0.03].

Although the correlations reported in Tables 4 and 5 are of a moderate order, extremely high levels of agreement are not expected in studies of concurrent validity. For example, correlations between the two alcohol consequences measures and TLFB drinking variables in Table 4 are similar to those reported comparing the Lifetime Drinking History (LDH) to the Alcohol Use Inventory (AUI) and to the MAST. The highest correlations were 0.53 and 0.50, respectively.[47,48] In another study, pretreatment alcohol intake correlated significantly but modestly ($r = 0.53$) with the Short Alcohol Dependence Data scale.[48a] Correlations between the TLFB and biochemical tests were higher than those reported between the LDH and similar tests, where the highest correlation was 0.28.[47,48] In addition, the association between frequency of drinking on the TLFB and abnormal liver function tests parallels findings from a recent study where subjects with high GGT test results had significantly more drinking days in the 30 d preceding treatment than subjects with normal GGT levels.[49]

What Instrument to Use

Although several drinking-assessment instruments exist, not all of them are suitable for every research study or clinical interview. TLFB is no exception, and this section is intended to provide guidance about when the use of it is indicated. As discussed elsewhere, before choosing a drinking instrument, a decision must be made about the level of precision and time frame that is desired.[19] The TLFB method will be briefly compared to three other major drinking-assessment methods: QF methods,[50–53] LDH,[54–56] and self-monitoring (SM).[57–59] For each of these methods, examples of when and why to use the method are briefly noted in Table 6 and are summarized below.

QF Methods

The first set of methods, discussed earlier in the chapter, are commonly referred to as quantity–frequency, or QF, methods. The two most frequently used QF methods include the one developed by Cahalan and his colleagues and the one used in a national survey of alcohol treatment programs by the Rand Corporation.[50–52] QF methods require people to report an average consumption pattern (e.g., How

Table 6
Measuring Drinking: What Instrument to Use and When

	Quantity–Frequency
Recommended use:	To obtain only total amount consumed or total number of drinking days in an interval
When to use:	If time is of a premium and little information is needed
	Lifetime Drinking History
Recommended use:	To obtain a lifetime or long-term picture of drinking
When to use:	To recall drinking over many years (e.g., from adolescence to adulthood) or in the distant past (e.g., from 1960 to 1980)
	Timeline Follow-Back
Recommended use:	To evaluate specific changes in drinking before and after treatment, or to get a picture of heavy and light drinking days
When to use:	If precision is wanted (e.g., percentage of days drinking at certain levels, weekend/weekday pattern changes), or to reflect risk days
	Self-Monitoring
Recommended use:	If slightly more accurate information is sought; however, the additional information would rarely be sufficient to alter a diagnosis.
When to use:	During treatment to provide ongoing feedback about substance use or urges. One major problem, however, is that not all clients comply with SM.

many days on average did you drink wine? On a day when you drank wine, how many glasses on average did you drink?). Although QF methods provide reliable information about total consumption and number of drinking days, as noted earlier, they do not identify atypical heavy drinking days, which are often associated with alcohol-related problems.[10] QF methods also cannot provide a picture of unpatterned fluctuations in drinking since their estimates are based on an assumption that the pattern of drinking is constant. QF methods, therefore, are recommended when all that is needed is the total amount consumed or the total number of drinking days in an interval, and when time is of a premium and information about atypical drinking is not required.

Lifetime Drinking History

The second drinking assessment method is the lifetime drinking history (LDH). This method was developed by Skinner and was recently refined and evaluated by Sobell and colleagues.[54–56] Using the LDH, lifetime drinking is recalled in discrete phases involving major changes in a person's average (i.e., QF) pattern. The LDH is not difficult to administer and takes about 20–30 min to complete. Since the LDH lacks precision for the most recent drinking period (e.g., in the last year), if information about a person's recent drinking is desired, LDH is not recommended. The LDH would be recommended when it is desired to produce a lifetime or long-term picture of a person's drinking. Examples include when seeking to assess drinking from adolescence to adulthood or for recalling drinking over a selected time period in the distant past.[60]

Timeline Follow-Back

Since considerable discussion has already occurred about the third drinking method, TLFB, only a few salient points will be reviewed. TLFB is a daily estimation method developed for the retrospective collection of data up to 12 mo prior to the date of the interview. Several memory aids are used to enhance recall. Depending on the time period assessed, it takes about 10–30 min to complete. TLFB is recommended when there is a desire to evaluate specific changes in drinking such as before and after treatment or when there is a need to have a fairly accurate picture of heavy and light drinking days. The precision that can be obtained includes the number of days drinking at certain levels as well as changes in drinking patterns over time.

Self-Monitoring

The fourth drinking assessment technique is self-monitoring (SM). SM logs or diaries can take a variety of forms and are designed to meet a client's needs.[19,61] Clients can be asked to routinely record various aspects of their target behavior (e.g., mood, urges, consequences, amount, frequency, and time of use). Compared to the other three methods, SM is subject to fewer recall problems when used appropriately (i.e., filled out shortly after the target behavior has occurred). However, there are two major problems with this method. Not all clients comply with the instructions; and unlike the three drinking-assessment measures reviewed above, SM cannot provide a retrospective assessment of drinking.[62] Although SM will provide slightly more accurate information than TLFB, the added information would not significantly change an evaluation of the severity of the person's drinking.[59,63]

QF–TLFB Comparison

Over the past several years, a number of studies have compared TLFB and various QF methods.[10,11,34,64] The next three figures (Figs. 3,4,5) present data comparing drinking reports from the same subjects using TLFB and two QF estimation formulae—the Cahalan QF and the QF method used in the Rand study.[11,34] In all three figures, the recall interval is 90 d, and data are from two subject populations (i.e., male and female normal drinker college students and male and female normal drinkers in the general population). Each table presents a different drinking variable: Fig. 3, mean drinks each drinking day; Fig. 4, total number of drinking days; and Fig. 5, total number of drinks consumed.

The two graphs in Fig. 3 plot subjects' TLFB reports of their mean drinks each drinking day by their QF category designation. In the top graph, college students were classified by QF criteria as being either heavy, moderate, or light drinkers; and in the bottom graph, normal drinkers in the general population were classified as either high or low consumers. Considering the bottom graph, the insensitivity of the QF method is indicated by the fact that it classified as high consumers those subjects whose TLFB reports for mean drinks each drinking day ranged from 1.6 to 8.2 drinks. It also is clear that the QF categories included considerable overlap in terms of mean levels of consumption. Similar problems are evident in the top graph with college students.

Figure 4 is similar to Fig. 3, except that it plots TLFB and QF reports by a different drinking variable: total number of days in the interval when drinking occurred. The bottom graph, for example, shows

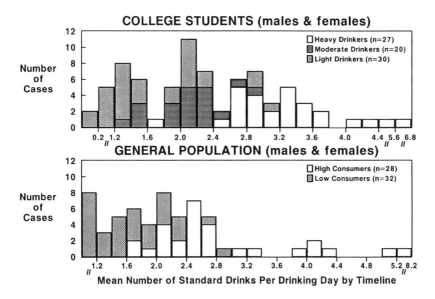

Fig. 3. Histograms showing the relationship between subjects' drinking classifications for two populations of drinkers based on their answers to two different QF questionnaires for mean number of standard drinks consumed each drinking day during the 90-d target interval as reported by the TLFB. Interval scaling is 0.00–0.19, 0.20–0.39, and so on.

that the total number of drinking days on the TLFB for subjects classified by the QF as high consumers ranged from a low of 16 to a maximum of 90 d. Similar findings are shown in Fig. 5 as in Fig. 3 and 4, except that the TLFB variable is total number of drinks consumed in the interval. Readers interested in further discussions about these graphs are referred to the original studies from which Fig. 3, 4, and 5 were compiled.[11,34]

The comparisons of the TLFB and QF methods in Fig. 3, 4, and 5 illustrate the importance of obtaining specific estimates of daily drinking rather than using estimation formulae. Collectively, the findings in these three figures suggest that the QF categories do not make meaningful distinctions in terms of specific dimensions of drinking. For each dimension, the QF categories overlap considerably. Although it could be argued that the QF categories are determined by multiple

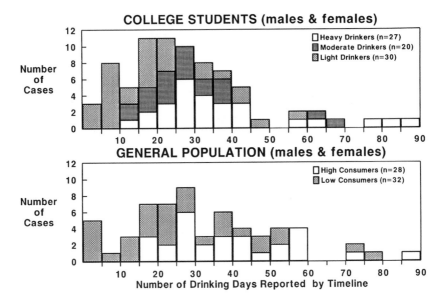

Fig. 4. Histograms showing the relationship between subjects' drinking classifications for two populations of drinkers based on their answers to two different QF questionnaires for total number of drinking days during the 90-d target interval as reported by the TLFB. Interval scaling is 0.00–4.99, 5.00–9.99, and so on.

factors and therefore, that one should not expect good correspondence with other measures on a single dimension, the evidence suggests that the overlap is an artifact of having subjects report what they perceive as their average drinking pattern.[10]

Computerized TLFB Program

Currently, TLFB is being developed for use as a computer-administered software program. While the program has been field tested, it is still being refined. The collection of data using the computer program is almost identical to its collection by paper-and-pencil method and in fact, has some decided advantages. Although computerized data collection has been more the exception than the rule in the alcohol treatment field, computers have been shown to interview patients about as accurately as professionals.[65,65a]

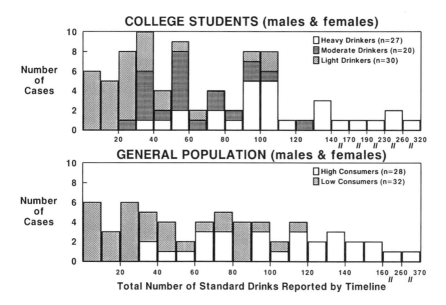

Fig. 5. Histograms showing the relationship between subjects' drinking classifications for two populations of drinkers based on their answers to two different QF questionnaires for total number of standard drinks consumed during the 90-d target interval as reported by the TLFB. Interval scaling is 0.00–9.99, 10.00–19.99, and so on.

Instructions for completing the calendar are part of the computer program and appear at the beginning before the person starts to complete the calendar. The program can collect data over intervals ranging up to 1 yr. The program also provides subjects with a country-specific, standard drink-conversion menu (e.g., Canada, US, Britain, and Australia). To facilitate recall, the computerized calendar lists major events and holidays. Subjects are also asked to list personal holidays and events. The program is flexible, allowing subjects to complete the calendar in more than one session. It also includes built-in consistency checks, e.g., checking that the maximum amount reported ever consumed on a single day is not exceeded by any single day in the interval. After the calendar is completed, the drinking data can be analyzed and graphed and immediate feedback can be provided to the subject. Appendix B contains a sample copy of the types of data that can be generated, at this time, by the computerized version of TLFB.

Summary and Conclusions

Perhaps owing to the variables it can generate as well as to its good psychometric characteristics, the TLFB method, since its inception, has been used in more than four dozen published studies and with populations as diverse as cancer patients, liver disease patients, arthritic patients, and prescription drug abusers.[35,66–69] Aside from being used with different drinker populations and different clinical disorders, several features of studies that have used TLFB with alcohol abusers are noteworthy:

1. The length of the intervals over which daily drinking data have been collected range from 3 to 18 mo;
2. Data have been collected using a single interview;
3. The interviews have been conducted in person, by phone, and by using both methods together; and
4. Several studies have used collaterals to check the validity of the subjects' self-reports.

Clearly, the TLFB technique has been shown to be the best psychometrically evaluated and field tested drinking-assessment instrument in the literature to date. However, as with all self-report–based techniques, there will always be some element of imprecision in the data collection. Thus, it is still important to develop further techniques and aids to help improve the accurate and complete recall of drinking. Also, studies that continue to provide concurrent validity for TLFB will add to the confidence with which it is used.

While the TLFB technique was developed to aid recall of drinking, its utility is not limited to assessing drinking behavior. For example, a recent study reported that the technique had been adapted for use with benzodiazepine users to assess benzodiazepine treatment use history.[70] It has also been suggested that TLFB may be useful in assessing binge eating and purging in patients with bulimia nervosa.[71]

In conclusion, although TLFB is clearly superior to QF methods, subjects' reports are nonetheless retrospective and therefore vulnerable to memory errors. Also, TLFB data are generally reliable, but as with all drinking-assessment methods, exact day-by-day precision cannot be assumed or necessarily expected. Since TLFB data represent subjects' best estimates of their drinking, a small amount of error is to be expected in the exact dating of drinking or reporting of exact amounts consumed. Overall, however, TLFB appears to provide a relatively accurate portrayal of drinking and has both clinical and research utility.

Acknowledgment

The views expressed in this paper are those of the authors and do not necessarily reflect those of the Addiction Research Foundation. Portions of this chapter—including all the figures, Tables 1 and 3–6, and Appendices A and B—are reprinted by permission from the Addiction Research Foundation.[35] The authors are indebted to Henzel Jupiter for his time and assistance in development of the TLFB computer software program.

References

[1]L. C. Sobell and M. B. Sobell (1981) Outcome criteria and the assessment of alcohol treatment efficacy, *Evaluation of the Alcoholic: Implications for Research, Theory and Treatment*, Research Monograph No. 5 (National Institute on Alcohol Abuse and Alcoholism, Rockville, MD), pp. 369–382.

[2]L. C. Sobell and M. B. Sobell (1982) Alcoholism treatment outcome evaluation methodology, *Prevention, Intervention and Treatment: Concerns and Models*, Alcohol and Health Monograph No. 3 (National Institute on Alcohol Abuse and Alcoholism, Rockville, MD), pp. 369–382.

[3]M. B. Sobell, S. Brochu, L. C. Sobell, J. Roy, and J. Stevens (1987) Alcohol treatment outcome evaluation methodology: State of the art 1980-1984. *Addict. Behav.* **12**, 113–128.

[4]R. Pearl (1926) *Alcohol and Longevity* (Knopf, New York).

[5]R. Room (1990) Measuring alcohol consumption in the United States: Methods and rationales, *Research Advances in Alcohol and Drug Problems, vol.10.* L. T. Kozlowski, H. M. Annis, H. D. Cappell, F. B. Glaser, M. S. Goodstadt, Y. Israel, H. Kalant, E. M. Sellers, and E. R. Vingilis, eds. (Plenum, New York), pp. 39–80.

[6]J. L. Fitzgerald and H. A. Mulford (1987) Self-report validity issues. *J. Stud. Alcohol* **48**, 207–211.

[7]P. J. Karhunen and A. Penttilä (1990) Validity of post-mortem alcohol reports. *Alcohol and Alcoholism* **25**, 25–32.

[8]G. Knupfer (1984) The risks of drunkenness (or ebrietas resurrecta): A comparison of frequent intoxication indices and of population sub-groups as to problem risks. *Brit. J. Addict.* **79**, 185–196.

[9]G. Pequignot, A. J. Tuyns, and J. L. Berta (1978) Ascitic cirrhosis in relation to alcohol consumption. *Intl. J. Epidemiol.* **7**, 113–120.

[10]L. C. Sobell, T. Cellucci, T. Nirenberg, and M. B. Sobell (1982) Do quantity-frequency data underestimate drinking-related health risks? *Am. J. Public Health* **72,** 823–828.

[11]L. C. Sobell, M. B. Sobell, G. I. Leo, and A. Cancilla (1988) Reliability of a timeline method: Assessing normal drinkers' reports of recent drinking and a comparative evaluation across several populations. *Brit. J. Addict.* **83,** 393–402.

[12]D. J. Armor and M. J. Polich (1982) Measurement of alcohol consumption, *Encyclopedic Handbook of Alcoholism.* E. M. Pattison and E. Kaufman, eds. (Gardner Press, New York), pp. 72–80.

[13]R. Room (1971) Survey vs. sales data for the U.S. *Drinking and Drug Practice Surveyor* **3,** 15,16.

[14]M. Russell, J. W. Welte, and G. M. Barnes (1991) Quantity-frequency measures of alcohol consumption: Beverage-specific vs. global questions. *Brit. J. Addict.* **86,** 409–417.

[15]M. B. Sobell and L. C. Sobell (1973) Individualized behavior therapy for alcoholics. *Behav. Ther.* **4,** 49–72.

[16]T. F. Babor, J. Brown, and F. K. Del Boca (1990) Validity of self-reports in applied research on addictive behaviors: Fact or fiction? *Addict. Behav.* **12,** 5–32.

[17]L. C. Sobell and M. B. Sobell (1990) Self-report issues in alcohol abuse: State of the art and future directions. *Behav. Assess.* **12,** 91–106.

[18]M. Phillips (1984) Sweat-patch testing detects inaccurate self-reports of alcohol consumption. *Alcoholism: Clin. Exp. Res.* **8,** 51–53.

[19]L. C. Sobell, M. B. Sobell, and T. D. Nirenberg (1988) Behavioral assessment and treatment planning with alcohol and drug abusers: A review with an emphasis on clinical application. *Clin. Psych. Rev.* **8,** 19–54.

[20]T. J. O'Farrell and S. A. Maisto (1987) The utility of self-report and biological measures of alcohol consumption in alcoholism treatment outcome studies. *Advances Behav. Res. Ther.* **9,** 91–125.

[21]L. C. Sobell (1978) Alcohol treatment outcome evaluation: Contributions from behavioral research, *Alcoholism: New Directions in Behavioral Research and Treatment.* P. E. Nathan, G. A. Marlatt, and T. Løberg eds. (Plenum, New York), pp. 255–269.

[22]L. C. Sobell (1978) Critique of alcoholism treatment evaluation, *Behavioral Approaches to Alcoholism.* G. A. Marlatt and P. E. Nathan eds. (Rutgers Center of Alcohol Studies, New Brunswick, NJ), pp. 166–182.

[23]J. M. Polich and C. T. Kaelber (1985) Sample surveys and the epidemiology of alcoholism, in *Alcohol Patterns and Problems.* M. A. Schuckit ed. (Rutgers, New Brunswick, NJ), pp. 43–77.

[24]E. M. Pattison, M. B. Sobell, and L. C. Sobell, eds. (1977) *Emerging Concepts of Alcohol Dependence.* (Springer, New York).

[25]D. L. Davies (1962) Normal drinking in recovered alcohol addicts. *Quarterly J. Stud. Alcohol* **23,** 94–104.

[26]D. L. Davies (1963) Normal drinking in recovered alcohol addicts (1962) *Quarterly J. Stud. Alcohol* **24**, 109–121, 321–332, 727–735.

[27]A. M. Cooper, M. B. Sobell, S. A. Maisto, and L. C. Sobell (1980) Criterion intervals for pretreatment drinking measures in treatment evaluation. *J. Stud. Alcohol* **41**, 1186–1195.

[28]A. M. Cooper, M. B. Sobell, L. C. Sobell, and S. A. Maisto (1981) Validity of alcoholics' self-reports: Duration data. *Intl. J. Addict.* **16**, 401–406.

[29]S. A. Maisto, L. C. Sobell, A. M. Cooper, and M. B. Sobell (1982) Comparison of two techniques to obtain retrospective reports of drinking behavior from alcohol abusers. *Addict. Behav.* **7**, 33–38.

[30]S. A. Maisto, L. C. Sobell, and M. B. Sobell (1979) Comparison of alcoholics' self-reports of drinking behavior with reports of collateral informants. *J. Consult. Clin. Psych.* **47**, 106–122.

[31]S. A. Maisto, M. B. Sobell, and L. C. Sobell (1982) Reliability of self-reports of low ethanol consumption by problem drinkers over 18 months of follow-up. *Drug Alcohol Depend.* **9**, 273–278.

[32]L. C. Sobell, S. A. Maisto, M. B. Sobell, and A. M. Cooper (1979) Reliability of alcohol abusers' self-reports of drinking behavior. *Behav. Res. Ther.* **17**, 157–160.

[33]M. B. Sobell, S. A. Maisto, L. C. Sobell, A. M. Cooper, T. Cooper, and B. Sanders (1980) Developing a prototype for evaluating alcohol treatment effectiveness, *Evaluating Alcohol and Drug Abuse Treatment Effectiveness: Recent Advances.* L. C. Sobell, M. B. Sobell, and E. Ward, eds. (Pergamon, New York), pp. 129–150.

[34]M. B. Sobell, L C. Sobell, F. Klajner, D. Pavan, and E. Basian (1986) The reliability of a timeline method for assessing normal drinker college students' recent drinking history: Utility for alcohol research. *Addict. Behav.* **11**, 149–162.

[35]L. C. Sobell and M. B. Sobell (1991) *Timeline Follow-back (TLFB) Manual and Instructions for How to Operate the TLFB Computer Software.* (Addiction Research Foundation, Toronto), (in press).

[36]M. B. Sobell, L. C. Sobell, and R. VanderSpek (1979) Relationships between clinical judgment, self-report and breath analysis measures of intoxication in alcoholics. *J. Consult. Clin. Psych.* **47**, 204–206.

[37]L. T. Midanik and A. M. Hines (1991) 'Unstandard' ways of answering standard questions: Protocol analysis in alcohol survey research. *Drug Alcohol Depend.* **27**, 245–252.

[38]J. R. McKay, T. J. O'Farrell, S. A. Maisto, G. J. Connors, and D. C. Funder (1989) Biases in relapse attributions made by alcoholics and their wives. *Addict. Behav.* **14**, 513–522.

[39]J. C. Nunnally (1978) *Psychometric Theory,* 2nd Ed. (McGraw–Hill, New York).

[40]S. A. Maisto, L. C. Sobell, M. B. Sobell, and B. Sanders (1985) Effects of outpatient treatment for problem drinkers. *Am J. Drug Alcohol Abuse* **11**, 131–149.

[41]H. A. Skinner and B. A. Allen (1982) Alcohol dependence syndrome: Measurement and validation. *J. Abnorm. Psych.* **91**, 199–209.

[42]M. L. Selzer, A. Vinokur, and L. van Rooijen, (1975) A self-administered Short Michigan Alcoholism Screening Test (SMAST). *J. Stud. Alcohol* **36**, 117–126.

[43]J. Chick, N. Kreitman, and M. Plant (1981) Mean cell volume and gamma-glutamyl-transpeptidase as markers of drinking in working men. *Lancet* **1**, 1249–1251.

[44]J. A. Cohen and M. Kaplan (1979) The SGOT/SGPT An indicator of alcoholic liver disease. *Digest. Dis. Sci.* **24**, 835–838.

[45]M. Y. Morgan (1980) Markers for detecting alcoholism, and monitoring for continued abuse. *Pharmacol. Biochem. Behav.* **13**, 1–8.

[46]M. Salaspuro (1986) Conventional and coming laboratory markers of alcoholism and heavy drinking. *Alcoholism: Clin. Exp. Res.* **10 (Suppl)**, 5s–10s.

[47]H. A. Skinner (1982) Development and validation of a lifetime alcohol consumption assessment procedure. (Addiction Research Foundation Substudy No. 1248. Toronto, Ontario).

[48]H. A. Skinner (1984) Instruments for assessing alcohol and drug problems. *Bull. Soc. Psycholog. Addict. Beh.* **3**, 21–33.

[48a]R. Davidson and D. Raistrick (1986) The validity of the short alcohol dependence data (SADD) questionnaire: A short self-report questionnaire for the assessment of alcohol dependence. *Brit. J. Addict.* **81**, 217–222.

[49]E. D. Richardson, P. F. Malloy, R. Longabaugh, J. Williams, N. Noel, and M. C. Beattie (1991) Liver functioning tests and neuropsychologic impairment in substance abusers. *Addict. Behav.* **16**, 51–55.

[50]D. Cahalan (1973) Drinking practices and problems: Research perspectives on remedial measures. *Public Affairs Rep.* **14**, 1–6.

[51]D. Cahalan and R. Room (1974) *Problem Drinking Among American Men* (Rutgers Center of Alcohol Studies, New Brunswick, NJ).

[52]J. M. Polich, D. J. Armor, and H. B. Braiker (1981) *The Course of Alcoholism: Four Years After Treatment* (Wiley, New York).

[53]R. Straus and S. D. Bacon (1953) *Drinking in College* (Yale University Press, New Haven, CT).

[54]H. A. Skinner (1978) The art of exploring predictor-criterion relationships. *Psych. Bull.* **85**, 327–337.

[55]H. A. Skinner and W. J. Sheu (1982) Reliability of alcohol use indices: The lifetime drinking history and the MAST. *J. Stud. Alcohol* **43**, 1157–1170.

[56]L. C. Sobell, M. B. Sobell, D. M. Riley, R. Schuller, D. S. Pavan, A. Cancilla, F. Klajner, and G. I. Leo (1988) The reliability of alcohol abusers' self-reports of drinking and life events that occurred in the distant past. *J. Stud. Alcohol* **49**, 225–232.

[57]R. O. Nelson and S. C. Hayes (1981) Theoretical explanations for reactivity in self-monitoring. *Behav. Modif.* **5**, 3–14.

[58]L. C. Sobell and M. B. Sobell (1973) A self-feedback technique to monitor drinking behavior in alcoholics. *Behav. Res. Ther.* **11**, 237–238.

[59]M. B. Sobell, J. Bogardis, R. Schuller, G. I. Leo, and L. C. Sobell (1989) Is self-monitoring of alcohol consumption reactive? *Behav. Assess.* **11**, 447–458.

[60]L. C. Sobell, M. B. Sobell, and T. Toneatto (1991) Recovery from alcohol problems without treatment, *Self-Control and Addictive Behaviors.* N. Heather, W. R. Miller, and J. Greeley, eds. (Pergamon, New York).

[61]T. D. Nirenberg, L. C. Sobell, S. Ersner-Hershfield, and A. J. Cellucci (1983) Can disulfiram precipitate urges to drink alcohol? *Addict. Behav.* **8**, 311–313.

[62]M. Sanchez-Craig and H. M. Annis (1982) 'Self-monitoring' and 'recall' measures of alcohol consumption: Convergent validity with biochemical indices of liver function. *Brit. J. Alcohol Alcoholism* **17**, 117–121.

[63]J. A. Samo, J. A. Tucker, and R. E. Vuchinich (1989) Agreement between self-monitoring, recall, and collateral observation measures of alcohol consumption in older adults. *Behav. Assess.* **11**, 391–409.

[64]S. A. Maisto and G. R. Caddy (1981) Self-control and addictive behavior: Present status and prospects. *Intl. J. Addict.* **16**, 109–133.

[65]M. W. Bernardt, O. J. Daniels, R. A. Blizard, and R. M. Murray (1989) Can a computer reliably elicit an alcohol history? *Brit. J. Addict.* **84**, 405–411.

[65a]D. R. Gavin, H. A. Skinner, and M. S. George (In press) Computerized approaches to alcohol assessment, in *Measuring Alcohol Consumption: Psychosocial and Biological Methods.* R. Z. Litten and J. Allen eds. (Humana Press, NJ).

[66]M. C. Shirley, D. L. Cookfair, L. C. Sobell, P. Steinig, P. A. Reese, G. J. Connors, and D. Heimbeck (1989) *The assessment of alcohol and high risk drinking situations in cancer patients.* Poster presented at the 23rd Annual Meeting of the Association for Advancement of Behavior Therapy, Washington, DC.

[67]U. A. Marbert, G. A. Stalber, G. Thiel, and L. Bianchi (1988) The influence of HLA antigens on progression of alcoholic liver disease. *Hepatogastroenterology* **35**, 65–68.

[68]A. Bradlow and A. G. Mowat (1985) Alcohol consumption in arthritic patients: Clinical and laboratory studies. *Ann. Rheum. Dis.* **44**, 163–168.

[69]E. M. Sellers, G. R. Somer, L. Sobell, and M. Sobell (1990) Discontinuation symptoms among persistent alprazolam users. *Clin. Pharmacol. Thera.* **47**, 168 (abstract).

[70]E. Schweizer, K. Rickels, W. G. Case, and D. J. Greenblatt (1990) Long-term therapeutic use of benzodiazepines. II. Effects of gradual taper. *Arch. Gen. Psychiatry* **47**, 908–915.

[71]G. T. Wilson (1987) Assessing treatment outcome in bulimia nervosa: A methodological note. *Intl. J. Eating Disord.* **6**, 339–348.

[72]G. J. Connors, D. W. Watson, and S. A. Maisto (1985) Influence of subject and interviewer characteristics on reliability of young adults' self-reports of drinking. *J. Psychopathol. Behav. Assess.* **7**, 365–374.

[73]S. A. Maisto, M. B. Sobell, A. M. Cooper, and L. C. Sobell (1979) Test-retest reliability of retrospective self-reports in three populations of alcohol abusers. *J. Behav. Assess.* **1**, 315–326.

[74]R. E. Vuchinich, J. A. Tucker, L. Harllee, S. Hoffman, and J. Schwartz (1985) *Reliability of reports of temporal patterning.* Poster presented at the annual meeting of the Association for the Advancement of Behavior Therapy, Houston, TX.

[75]T. J. O'Farrell, H. S. G. Cutter, R. D. Bayog, G. Dentch, and J. Fortgang (1984) Correspondence between one-year retrospective reports of pretreatment drinking by alcoholics and their wives. *Behav. Assess.* **6**, 263–274.

[76]B. S. McCrady, N. E. Noel, D. B. Abrams, R. L. Stout, H. F. Nelson, and W. M. Hay (1986) Comparative effectiveness of three types of spouse involvement in outpatient behavioral alcoholism treatment. *J. Stud. Alcohol* **47**, 459–467.

[77]B. S. McCrady, R. Stout, N. Noel, D. Abrams, and H. F. Nelson (1991) Effectiveness of three types of spouse involved behavioral alcoholism treatment: Outcomes 18 months after treatment. *Brit. J. Addict.* **86**, 1415–1424.

[78]A. Zweben (1986) Problem drinking and marital adjustment. *J. Stud. Alcohol* **47**, 167–172.

[79]A. Zweben, S. Pearlman, and S. Li (1988) A comparison of brief advice and conjoint therapy in the treatment of alcohol abuse: The results of the marital systems study. *British J. Addict.* **83**, 899–916.

[80]L. C. Sobell, M. B. Sobell, S. A. Maisto, and W. Fain (1983) Alcohol and drug use by alcohol and drug abusers when incarcerated: Clinical and research implications. *Addict. Behav.* **8**, 89–92.

Appendix A

INSTRUCTIONS FOR COMPLETING
THE TIMELINE DRINKING CALENDAR

Using the attached calendar, we would like you to reconstruct your drinking for the time period indicated on the calendar. This is not a difficult task, especially when you use the calendar for reference. We have found calendars useful in helping people recall their drinking. The following are instructions and tips for completing the calendar:

INSTRUCTIONS

❶ It is important that for *each* day listed on the calendar, there is a number indicating the number drinks you consumed. In reporting your total daily consumption, we would like you to report it in **STANDARD DRINKS** (use the standard drink equivalent card).

❷ On the days that you did not drink any alcoholic beverages mark those days with a "0".

❸ On the days that you did consume a beverage containing alcohol, write in the total number of **Standard Drinks** that you drank on those days. This includes days of combined beverage use. For example, if you drank a glass of wine with dinner and a drink containing 1-1/2 oz. of hard liquor after dinner, you would count that as 2 standard drinks for that day. **The important thing is to make sure that something is filled in for each day.**

❹ In filling out the calendar, we would like you to be as accurate as possible. However, if you cannot recall whether you consumed an alcoholic beverage on Monday or Thursday of a certain week, or whether it was the week of November 9th or the week of November 16th, **give it your best shot!**

The purpose of the calendar is to get as accurate a picture of what your drinking has been like for the indicated time period in terms of number of drinking days and number of drinks per day.

HELPFUL HINTS

☛If you have an **appointment book** or a **daily diary** available, you can **use** it to help you recall your drinking,

☛As you will notice, **standard holiday days** are **marked on the calendar** to help your recall; you can also write in special holidays such as birthdays, vacations, celebrations.

☛Some people have **regular drinking patterns** and this can help them in filling out the calendar. For example, you may have a **weekend/weekday change** in your drinking or your drinking may be different depending on the season, or whether you are on holidays or business trips.

ADDITIONAL SUGGESTIONS FOR ASKING FOR TIMELINE FOLLOW-BACK INFORMATION

Incarcerations or Confinements

When interviewing alcohol abusers, if they mention hospitalizations, treatment stays, or jail periods that occurred in the recall period, it is often easy to start with those events as they are discrete and time bound.

For example, the interviewer can ask: "You said that you were in a hospital sometime in the last year. What were the dates you were there? Did you have anything to drink during this time?"

NOTE: Stays in jails, hospitals, or residential treatment programs do not preclude frequent drug or alcohol use (*see* Sobell, Sobell, Maisto, & Fain, 1983).[80] Thus, to obtain accurate data, it is important to assess alcohol use during periods of incarcerations. On the calendar, these days are coded as both incarcerated and an amount consumed is listed.

Probing Extended Abstinent or Drinking Periods

"Were there any times in this recall period when you had **nothing at all to drink**, not even a drop of alcohol?"

"What was the longest period of total abstinence during this time?"

"What was the next longest period of total abstinence?"

"What was the longest number of continuous days in a row you were drinking during this period?" (determine dates and amounts of alcohol consumed on each day)

"What was the next longest period of continuous drinking days?"

Other Events

Sometimes when being interviewed people will report not drinking during a particular period. For example, a person may say "I wasn't working during October, so I had no money to drink, but when I returned to work in May I started drinking." These time periods can be listed on the calendar and then questions can be directed to the time periods around such events.

Sometimes people drink routinely after or at particular events (e.g., bowling on Mondays; hockey on Thursdays; playing pool). To this end, the interviewer can specifically ask the person

"Were there any times or events where you almost always drank alcohol? For example, ... ".

Completing the Calendar in a Flexible Manner

Wherever people feel comfortable in filling out the calendar can be the starting date. People can go forwards or backwards from the interview date or jump around from month to month.

If a person has trouble recalling his/her drinking, try working back from when the person entered treatment. Say "What about this past month, what was your drinking like then?" The most recent months should be the most familiar, and the person might find it easier to reflect upon those periods. Ask questions about special occasions (birthdays, weddings) and use these days as anchors or reference points to help the person better recall his/her drinking.

Appendix B

Sample Data Output from TLFB Computer Program

Table A
Subject 066666, Session 1, Running from February 27, 1991 to April 2, 1991

January	July
February	August
7 8	
March	September
5 6 7 7 0 0 0 0 0 0 0 0 7 4 5 5 21 2 4 0 0 0 0 33 4 7 8 8 9 10 20	
April	October
3 3	
May	November
June	December

Table B
Histogram of Percentage of Days of Collected Data*

Data		%
0		34.3
1		0.0
2		2.9
3		5.7
4		8.6
5		8.6
6		2.9
7		14.3
8		8.6
9		2.9
10		2.9
11		0.0
12		0.0
13		0.0
14		0.0
15		0.0
16		0.0
17		0.0
18		0.0
19		0.0
20		2.9
21		2.9
22		0.0
23		0.0
24		0.0
25		0.0
26		0.0
27		0.0
28		0.0
29		0.0
30		0.0
31		0.0
32		0.0
33		2.9

*Subject 066666, session running for 35 d. Form: 4. Age: 29. BAC: 0. The maximum number of data collected is 33 for March 24. The maximum number of continuous days of 0 data is 8 from March 5 to March 12. The maximum number of continuous days of non-0 data is 10 from March 24 to April 2.

Table C
Subject 066666, Session 1, Running for 35 d

Monthly data		Daily summary data						
January	0	Sun.	Mon.	Tue.	Wed.	Thu.	Fri.	Sat.
February	15	81	16	14	22	20	19	21
March	172							
April	6							
May	0							
June	0							
July	0							
August	0							
September	0							
October	0							
November	0	Administered Tuesday April 2, 1991						
December	0	Start Date: Wednesday February 27, 1991						
Total	193	Finish Date: Tuesday April 2, 1991						

Table D
Subject 066666, Session 1, Running from February 27, 1991 to April 2, 1991*

Category	Abstinent	Low ethanol consumption	High ethanol consumption	Very high ethanol consumption
From	—	1	5	10
To	0	4	9	- >
For 30 days				
From March 4	12	6	8	4
To April 2	34%	17%	23%	11%
For 60 days				
From Feb. 2	12	6	13	4
To April 2	34%	17%	37%	11%

*Percentage of days in which drinking occurred: 65.7, consuming 8.4 drinks/drinking day. Early-morning drinking has occurred.

Using Subject and Collateral Reports to Measure Alcohol Consumption

Stephen A. Maisto and Gerard J. Connors

Introduction

In this chapter, we will review and discuss the literature on the use of collateral (C) and subject (S) reports in the measurement of a subject's alcohol consumption. The word *collateral* has several definitions, but here it should be viewed to mean accompanying or concomitant. As regards assessment of alcohol consumption, C reports provide a second measure of a subject's drinking. In practice, the subject is asked to report about his or her drinking during a given time period (one measure), and the collateral is asked to report on the same information (second measure). Collaterals may have any relationship to subjects (spouse, friend, supervisor, and so forth); the only requirement is that the person has some basis, such as opportunity to observe, for reporting about a subject's drinking.

The chapter will begin with background on the use of C reports in measuring drinking. We then will discuss findings on the use of C reports from the 1986 National Institute on Alcohol Abuse and Alcoholism (NIAAA) workshop on the reliability and validity of self-reports of drinking. Following that, we will present a review of

From: *Measuring Alcohol Consumption*
Eds.: R. Litten and J. Allen ©1992 The Humana Press Inc.

empirical studies of the use of C reports that have appeared since 1986. This empirical base helped to identify six research issues about the use of C reports. We will discuss each and then draw several conclusions about the use of C reports in measuring drinking.

Importance of C Reports

Collateral reports are important in the measurement of a person's alcohol consumption because there remains no flawless way to obtain measurements of many alcohol consumption variables. This problem, popularly called a lack of a "gold standard," is especially acute in obtaining measurements of a person's past drinking. In obtaining such retrospective measures, researchers and clinicians alike have to rely primarily on self-report, so that the C report becomes an important second measure. We will say more later in this chapter about how C reports traditionally have been conceptualized.

Collateral reports have several properties that make them valuable second measures of a subject's drinking. First, as with S reports, C reports are highly flexible in the content of the measure and the time period that the measure covers. Therefore, C reports can be second measures of a person's drinking that theoretically cover any pattern of drinking over any time period. In practice, this has meant asking the C to report on a subject's drinking for a time as long as 5 or 6 yr ago to the last 24 h. As we alluded to above, the flexibility of C reports becomes more important the more remote in time we are inquiring about the subject's drinking. With current technology, when discussing drinking patterns of more than a month ago, we do not have strong alternatives to measurement other than S and C reports.

Another reason for the importance of C reports is that they are relatively inexpensive second measures of drinking. Again, as with S reports, the cost of C reports is only as high as the staff time it takes to contact and interview the C and to provide payment, if any, to the C for completing an interview. Such costs compare quite favorably to other measures of alcohol consumption, such as those involving urine or blood testing. In addition, C reports are low-burden measures. Asking a person to report on another individual's drinking is a simple, relatively noninvasive procedure. As we will show in the review of empirical studies of the use of C reports, in general, subject and collateral show few reservations in consenting to collateral interviews about the subject's drinking.

Traditional View of Validating Self-Report

Collateral reports emerged as an important way to measure drinking for the same reasons other additional measures (e.g., breath tests, urine tests, and official records) emerged: For different reasons, an S report about drinking and related behaviors may not be accurate. One major reason is the context of the assessment, which may have associated incentives for an S to underreport (appear to function better than actuality) or to overreport (appear to be functioning worse) their alcohol consumption. Context factors will be discussed in detail in Thomas F. Babor's and other chapters, so we will comment only briefly here.

By far, most of the discussion has focused on incentives for a subject to underreport his or her drinking. Many such contexts, relating to home, job, or legal circumstances, can be imagined in which subject might see it to be in his or her interests to appear to be drinking less, or not at all, or to be experiencing no problems related to drinking. In any case, researchers and clinicians traditionally have assumed that subjects are aware of incentives to appear to be functioning well and that they are acted upon.

Much less discussed are incentives to overreport alcohol consumption or related consequences. Here, subjects view it as worth their while to appear to be drinking more and having more alcohol problems. One example of a context that had this incentive is a study reported by Myers (1983).[1] This study involved criminal offenders' reports of alcohol consumption at the time they committed the crime for which they were currently imprisoned. In this context, there is an incentive for subjects to over-report their drinking because it might be seen as a mitigating circumstance of their criminal behavior.

A major concern in the alcohol- and drug-treatment field that has influenced the use of C reports is denial. That is, individuals having problems with alcohol or other drugs are assumed to be in denial about their substance use and associated consequences. As a result, when asked about drinking and drug use, or about any problems resulting from such use, the tendency would be for such individuals to underreport. Denial differs from incentives to appear better functioning (underreport) in that it is thought to reflect largely an unconscious rather than a conscious action by the individual.

Role of C Reports and Shaping Research Questions

The belief that S reports may not be accurate resulted in the search for additional measures to corroborate them. Collateral reports, as well as other measures, often were viewed as standards or criteria of how truthful or accurate the S report was. In the literature on C reports, this resulted in researchers' use of three major indices as bases of deciding whether a self-report was accurate: percent agreement, mean difference in the level of a variable, and amount and direction of discrepancy between S and C reports.

Percent agreement is a statistic that is used with variables measured on a nominal or categorical scale. There are many such measures of alcohol consumption. Examples are whether an individual has been abstinent from alcohol during some time period, classification of a person according to quantity and frequency of alcohol use,[2] and diagnosis (e.g., alcoholic or not).

To illustrate how percent agreement is computed for S and C reports, we will use the example of abstinence from alcohol. Assume that a clinician is interested in evaluating how patients are functioning after their completion of an alcohol rehabilitation program. Part of the evaluation design includes contacting patients 6 mo after treatment and asking them, among other questions, whether they have had any alcohol at all during the time period in question. If the patient says no, then he or she is scored as abstinent, if yes, nonabstinent.

As a second measure, the patient's spouse is asked the same question about the patient, and his or her response is similarly scored. The percentage of patient–spouse pairs in a group whose responses fall in the same category (abstinent or nonabstinent) results in the percent agreement statistic, which ranges from 0 to 100%. Percent agreement may vary in its computation according to the research design and the measure used, but essentially, it is designed to reflect how frequently two people agree on a category of response. Also note that the statistic may be applied to measures with more than two categories of response, such as the typical quantity–frequency (QF) of drinking classification.

Another index that frequently has been used is the mean difference between S and C reports in the level of a variable. This index is used when a variable is measured on a continuous scale. Measures of alcohol consumption of this type include, for example, number of days of drinking a particular quantity of alcohol, the amount of alcohol consumed on each drinking occasion, and the number of days of abstinence from alcohol. The typical presentation in research or clinical reports is to show the mean value of the measure based on S reports and the value of that same measure based on C reports. If the two mean values are not statistically different, then agreement between the two is judged to be acceptably high.

The final major index is related to reported mean difference in level of a variable. It is based on the discrepancy between S and C reports and the direction of discrepancy. Direction usually is concerned with whether the subject is reporting more or less of an event (e.g., number of drinking days, or number of drunk days, number of days drinking no alcohol at all) than does the collateral. In a sense, discrepancy analyses supplement analyses of mean difference in level and give an idea of variance in the self-report data along with mean level. Discrepancy also addresses the question of whether alcoholics are in denial about how much alcohol they drink and the consequences of such drinking.

Collateral reports traditionally have been viewed as a way to validate what a subject says about his or her alcohol consumption. That perspective influenced how S and C report data were presented, with the indices of percent agreement, mean difference in level of a variable, and S and C report discrepancies most commonly used to summarize self-report data. This is evident in reports of empirical studies on the use of S and C reports in measuring a subject's alcohol consumption.

The 1986 Panel Discussion
on the Reliability and Validity of Self-Reports

Because of the continued high rate of use of S reports as measures of drinking, the NIAAA held a panel discussion on the use of self-report measures of alcohol consumption. The discussion centered on the reliability and validity of S reports. Alan Marlatt and colleagues presented an excellent integration and summary of the empirical

literature up to 1986 on self-report measures of alcohol consumption.[3]
Table 1 shows the summary of the data on the use of C reports as a way
to validate S reports. It is useful to review Table 1 to provide a reference for the review of later empirical research on S and C reports that
will be summarized later in this chapter.

There are several notes about Table 1 that are important for interpreting the data contained in it. The numbers in parentheses designate
references that are data sources; we will include them as an appendix
to this chapter. Another point is that only two types of indices are
summarized in the table: percent agreement and Pearson correlation
coefficients. We explained the percent agreement index earlier. The
correlation coefficient is used for variables measured on a continuous
scale and reflects the simple linear relationship between S and C
report measures of a variable.

The data summarized in Table 1 are the foundation of a number
of important conclusions about the use of C reports. Following is a summary of the conclusions presented by Marlatt et al.[3]

1. The magnitude of the correlations varies, but overall, they are
   positive, in the moderate–good range, and statistically significant in
   all studies;
2. Consistency in the number of abstinent days is excellent in both
   outpatient and inpatient populations. With some exceptions, the
   number of days of light and heavy drinking show less correspondence;
3. Measures of the amount of alcohol consumed generally show lower
   correlations than measures of frequency or number of days of drinking. This was explained by the greater difficulty for the collateral to
   observe amount compared to the frequency of drinking;
4. Categorical measures of QF and global ratings of drinking pattern
   show good subject–collateral agreement;
5. Pre- or postassessment time does not affect consistency;
6. Subjects were informed in all but one study that a collateral would be
   interviewed about the subject's drinking. However, subject–collateral agreement in that one study was comparable to that observed in
   the studies where the subject was informed;
7. Discrepancy analyses showed no consistent tendency for the subject
   to overreport or underreport drinking; and
8. Spouses of subjects are the collaterals used most frequently, but there
   is a range of relationships between subject and collateral in all studies
   except those involving couples, such as in studies of marital therapy.

A number of studies on the use of C reports as measures of alcohol consumption and related behaviors were published up to 1986. The data, based almost entirely on clinical populations, provide overall support for the use of C reports as measures of alcohol consumption. We next turn to more recent studies of the use of C reports and compare the conclusions drawn from these data to the Marlatt et al. conclusions.[3]

Empirical Findings Since 1986 on the Use of C Reports

Our review of the empirical literature on S and C reports is summarized in Table 2. It is important to make some comments about the table. The references cited include published studies or paper presentations; data that have not appeared in either context were excluded from the review. Another point is that the nine studies presented in Table 2 focus on S–C report consistency. Therefore, S and C report data that were only secondarily reported, say, as part of a treatment outcome study, were not included. Finally, the table includes quantitative entries of percent agreement and Pearson correlation, to be consistent with the organization of Table 1. Thus, analyses involving indices such as mean levels or amount of discrepancy are not in Table 2.

Table 2 shows that the magnitude of the correlations and percent agreement findings generally are consistent with the older data presented in Table 1. Of significance also is that the data in Table 2 were collected from a range of clinical and nonclinical subject populations and, like the older research, cover a wide range of measures.

A few aspects of the data are important to discuss. The Fuller, Lee, and Gordis (1988) study showed poor agreement (47%) for the measure of number of drinking days over 1 yr.[4] We hypothesize that their finding likely was due to the recall period covered: One yr is a long time for exact agreement on a relatively specific event like number of drinking days. Ridley and Kordinak's (1988) finding of a 0.34 correlation between S and C reports on the Alcoholism Severity Rating Scale (ASRS) was probably the result of the qualities of that scale.[5] One feature that would tend to result in less consistency is that some of the items do not refer to events (e.g., did a person drink on that day?) but require the respondent to make judgments (e.g., To what

Table 1
Validity of Self-Reported Drinking and Related Behavior Using C Reports as Criteria[*,3]

	Measure				
Population	Days abstinent	Days drinking	Days light drinking[†]	Days heavy drinking[†]	Amount consumed
Inpatient alcoholics pretreatment	0.88 [37]	0.91 [37]	0.38 [37]	0.59 [37]	0.46 73–100%[34–38]
Inpatient alcoholics at followup	0.81 [30]	0.66 [28]	0.49 [30]	0.82 [30]	0.46 [28]
Outpatient problem drinkers at pretreatment	—	—	—	—	0.41–0.64 [35]
Outpatient alcoholics and problem drinkers at followup	0.84–0.86 [31]	—	0.94–0.95 [31]	0.56–0.68 [31]	0.42–0.92 [35]
Alcoholics not in treatment	—	—	—	—	—
Prisoners	—	—	—	—	0.51–0.68 [36]
College students	—	—	—	—	—

Population	Measure					
	QF index	Rating of pattern of drinking[‡]	Rating of functioning[§]	Alcohol-related symptoms and consequences[‖]	Hospital days alcohol-related	Jail days alcohol-related
Inpatient alcoholics pretreatment	0.52 56%[33]	77–86% —[33]	0.69 —[37]	0.65 67–76%[32,33]	0.98 —[37]	0.99 —[37]
Inpatient alcoholics at followup	0.34–0.59 84–96%[33]	0.57–0.76 73–80%[40,41]	0.84 80–100%[26,30,40]	0.50–0.61 84–88%[28,33]	0.97 —[30]	0.46 —[30]
Outpatient problem drinkers at pretreatment	—	—	—	—	—	—
Outpatient alcoholics and problem drinkers at followup	—	—	—	—	—	—
Alcoholics not in treatment	—	—	—	0.70 79%[29]	—	—
Prisoners	—	—	—	66–99%[27]	—	—
College students	0.68[39]	0.71[39]	—	0.79[39]	—	—

Note: Superscript characters are reference numbers.

*Decimal numbers are statistically significant ($p < 0.05$) correlation coefficients. Percentage figures represent agreement between S and C reports for categorical measures. Ranges of coefficients and percentages represent variation in results across studies, measures, or different assessment intervals. [†]Two studies used 3 oz of absolute alcohol a day as the cutoff for light and heavy drinking; one study[30] used 1-1/2 oz of absolute alcohol as the cutoff. [‡]For example, normal vs abnormal or abstinent vs controlled vs uncontrolled. [§]Includes ratings of improved vs unimproved following treatment, composite measures of days functioning well, and ratings of health, employment, family trouble, and so on. [‖]Includes alcoholism severity measures such as MAST and Alcohol Impairment Scale, as well as idiosyncratic measures of alcoholism-related symptoms, behaviors, and consequences.

Table 2
Empirical Findings Since 1986 on Subject–Collateral Consistency

Reference	Population	Measures and results
42	Inpatient alcoholics, pretreatment	SAAST, 95%
4	Inpatient alcoholics, 1-year followup	No. of drinking days in last 2 mo.,74% Abstinence in last 2 mo. 92–93%; no. of drinking days in last year, 47%
6	Natural recoverers nonresolved controls	No. of negative consequences, 0.77† Mean no of drinking days/mo., avge/max., 0.52†/0.16; mean percent morning drinking, 0.64†; Mls. alcohol/drinking day, avg/max., 0.06/0.31; mean percent of beverage consumption consisting of beer/wine/spirits 0.82†/0.71†/0.78†
43	Outpatient and inpatient alcoholics, combined, pretreatment	SAAST, 0.63†; annual absolute alcohol intake, 0.62†
5	Alcoholics post detox/controls	Quantitative Inventory for Alcohol Disorders, 0.51†/0.69† MAST, 0.53†/0.79†; Alcoholism Severity Rating Scale, .34/0.99†
24	General population of adults, husbands, wives	Model quantity of drinking, 54% (husbands' drinking); 67% (wives' drinking); complaints about drinking, 76%
44	Individuals "naturally recovered" from alcohol problems	Date of resolution, (month, year), 81%
20	Nonclinical sample of the elderly, range of drinking patterns	Classification measures‡, 0.63†–0.84† Proportion of days drinking 4–12 wk, 0.89†–0.93†
45	Inpatient alcoholics, 1–5 yr followup	Sobriety, whole time, 94%; alcohol use, 86%

Note: Percentages refer to percent agreement between subject/collateral pairs; decimal numbers are correlation coefficients.
* $p < 0.05$. † $p < 0.01$. ‡Drinking Practices Questionnaire[2], the Oates and McCoy Scale, and the Habitual Alcohol Use Questionnaire.

extent has the patient lost the ability to effectively abstain or regulate his or her drinking?). Furthermore, some items on the ASRS are vague in definition or refer to events that a collateral would have difficulty observing. In the Gladsjo, Tucker, Hawkins, and Vuchinich (1991) report, poor consistency on the maximum measures may have been caused by, as the authors suggest, the measure's relatively high specificity and resulting greater difficulty for free recall.[6] Importantly, these data on the maximum measures are similar to what has been found with clinical samples in use of the Lifetime Drinking History (LDH) measure, which Gladsjo et al. used in obtaining their maximum measures. Finally, the lower consistency found in both average and maximum quantity measures likely was the result of the difficulty a collateral would have in observing such events.

Comment on the Methodology Used in Recent Research

Before drawing conclusions on the use of C reports based on recent research, we will discuss the methodological quality of the studies. Those methods affect interpretation of the findings. Skinner (1984) reviewed procedures that research has shown tend to increase the reliability of an S report of his or her drinking and related behaviors.[7] The procedures were derived from findings on clinical samples but would seem to apply, where relevant, to nonclinical samples as well. The seven procedures that Skinner included are as follows:

1. Subjects are interviewed when they are alcohol- and drug-free;
2. Subjects are stable and showing no major symptoms from alcohol or other drug withdrawal or psychiatric illness;
3. Questionnaire items are carefully developed and structured;
4. Subjects are aware that their responses will be checked against other sources (e.g., collaterals, official records, or breath tests);
5. There is a good rapport between subject and interviewer;
6. There is no apparent reason for the subject to overreport or underreport his or her drinking; and
7. Subjects are assured of the confidentiality of their responses.

The more recent literature that we reviewed on S–C report consistency showed an overall excellent match between their procedures and Skinner's criteria. Therefore, the literature we reviewed followed methods that would tend to be more sensitive to finding S–C report consistency, at least from the subject side of the interview. Following more methodologically sound practices increases the confidence in the conclusions drawn from a body of data. We reached several conclusions on the use of C reports, based on empirical study of the question since 1986.

Conclusions About the Use of C Reports

Our conclusions relate both to the accuracy of an S report of drinking and to using a C report as a measure of a subject's drinking. Except for the Fuller et al. study, the data consistently suggest that subjects give accurate reports about their drinking. This conclusion spans research done with various clinical and nonclinical subject populations over a range of assessment intervals (e.g., 6 mo pretreatment, three months posttreatment followup), a range of categorical and continuous measures, and the use of collaterals that had a variety of relationships (e.g., spouse, friend, or relative) with subjects. Therefore, in view of the methodological basis of the research and the different populations it includes, in general, there can be considerable confidence in the accuracy of a subject's report of his or her drinking and in the use of C reports as a measure of drinking.

This conclusion is based on two indices of consistency: percent agreement and Pearson correlations. However, the papers cited in Table 2 often included mean differences in S and C reports and discrepancy analyses when continuous measures were reported. These two indices show that typically, there were not significant differences between S and C reports in the average value of a measure. Furthermore, when discrepancies occurred, there was no systematic trend for the S report to be higher or lower than the C report. These findings, of course, are consistent with the percent agreement and the correlation data.

Although in general the data showed good S–C report consistency, there were some exceptions. We referred to some of these earlier. It seems that poorer consistency is associated with more molecular, less well-defined, and harder-to-observe (for the collateral) behaviors. Also, the longer the referent recall period, the poorer the consistency that can be expected.

Consistent with earlier data, in recent research, we could find no pattern of predicting S–C report agreement as a function of the frequency of contact between subjects and collaterals, the relationship of the subject to the collateral, subject demographic characteristics, or subject drinking and related factors. We should note that no study investigated all of the above variables in, say, a multiple regression model. However, the lack of power of the variables to account for variance in S–C report agreement is surprising.

A final point about the more recent data is that they not only are consistent with previous research but also extend it through inclusion of nonclinical populations. In particular, samples of the elderly and natural recoverers further increased the generalizability of the conclusion that C reports can be useful measures of drinking.

Research Issues

In this section, we will discuss six research questions that we have identified in our integration of the data on the use of C reports. The research issues relate both to methodological features of the studies as well as to future lines of research in pursuit of advancing knowledge about and increasing the sensitivity of C reports. Other papers relevant to developing a research agenda on the use of C reports include the chapter by Babor in this volume and papers by Maisto, McKay, and Connors (1990), Midanik (1982, 1988, 1989), O'Farrell and Maisto (1987), Polich (1982), and Sobell and Sobell (1986).[8-14]

Problem of Chance Agreement

To some degree, the conclusion that C reports are a useful measure of drinking may be a false-positive. In this respect, the data available on the use of C reports is largely uncorrected for chance agreement. Therefore, S and C reports may be consistent in their reports, to some extent, merely by chance rather than by the accuracy of their reports. Accordingly, a valid interpretation of the data requires taking chance agreement into account in any analysis of S–C report consistency.

The studies that we and others have reviewed on the use of C reports have not included analyses that correct for chance agreement. This is unfortunate, since statistical methods have been available to do that. In this regard, the statistic kappa can be used to evaluate the degree of agreement between two categorical variables, corrected for chance, and the intraclass correlation coefficient (ICC) does the same

for continuous variables.[15-17] In our review of the research, we calculated kappa where possible for several studies that included percent
agreement data. In these studies, the conclusions drawn from the kappa
and percent agreement statistics were the same. Conclusions based on
kappa statistics, however, give more confidence in their validity. Therefore, it would be a methodological advance in the study of S–C report
agreement to provide kappa and ICC statistics where relevant.

What Is Acceptable Validity or Consistency?

In the literature on self-reports of alcohol consumption, researchers freely use the term validity but have no consistent standard to
determine what level of validity is acceptable. In practice, the question of acceptable validity usually is evaluated by the statistical significance of some index such as a correlation; there is no statistical
test for evaluating straight percent agreement data, but kappa
and ICC statistics can be assessed for significance. However, use of
only significance tests to determine acceptability of validity
still leaves the problem of lack of a generally useful method to evaluate validity. It would seem that *acceptable* needs to be defined in
terms of what the measure will be used for. Along these lines, it is
useful to consider the approach that psychometricians have used in
addressing this problem.

In traditional psychometric theory, validity refers to a test's
actually measuring what it was designed to measure (e.g., some psychological construct, such as self-esteem).[18] This definition of validity most closely relates here to the accuracy of an S report of alcohol
consumption: To what extent does a person's report of drinking
reflect the drinking that actually occurred? It can be seen that in the
self-report literature, when researchers and clinicians talk of the
validity of such reports, the question really concerns their accuracy.

Psychometricians operationalize the validity of a test by calcu-
lating the simple Pearson correlation between that test and some other
measure of theoretical (e.g., one that measures the same construct) or
practical (e.g., using the test to predict the score on some criterion
measure) interest. A validity coefficient of 0.80—considered high for
psychological measures—reduces error in predicting the exact score
on some other relevant measure by only 40%. Reduction of error in

prediction by 40% may not seem substantial, but it is of value. There is practical utility because usually the goal is to predict a range of values on the criterion measure, not an exact value. In any case, determining what is good or acceptable validity is a function of what the test will be used for.

This discussion of the psychometric concept of validity raises a few important points for the S–C report literature: Overall, the research shows that by the 0.80 standard, the degree of S–C report consistency is good to excellent. More importantly, this discussion shows that alcohol researchers have not spent enough time specifying the use of the drinking data they are evaluating. If pinpoint accuracy of any one individual's drinking behavior is required, then 0.80 agreement between S–C reports may not engender much enthusiasm; but, if estimates of drinking behavior based on group averages are required, then a 0.80 relationship between S and C reports may be deemed very good. It is the researcher's and clinician's task to specify what the drinking measure will be used for and, consequently, what an acceptable level of S–C report consistency is.

What Measure Should Be Used?

The S–C report empirical data showed that degree of consistency tends to vary with several variables, including how precise (sensitivity of the measure to small differences) the measure is,[19] how difficult it is for the collateral to observe the behavior that the measure is supposed to reflect (difficulty tends to be positively related to precision), and measurement intervals (shorter intervals are associated with greater consistency). Despite the variety of measures that have been evaluated for S–C report consistency, typically there was no justification for the measures that were used in a study.

This point relates to our comments above on defining an acceptable level of validity. The first consideration is whether the self-report measure of alcohol consumption or related behaviors fits the research or clinical needs. Along these lines, what research or clinical questions are being addressed? Will data analyses be on a group or individual level (e.g., 20). When these conceptual questions are answered, the goal is to achieve as much precision as possible in a self-report measure and still be able to defend its accuracy. Of course, it is highly desirable to have precise measures, but a more global measurement that can be defended as accurate is preferred to a precise measurement that cannot.

Framing the Research Question

A premise that has run through much of the literature on the validity of self-reports of alcohol consumption, especially the older research, is that a C report is the standard against which an S report is compared (Table 1). However, it would appear that this is not a viewpoint that is most likely to advance knowledge about S reports. This approach essentially closes off research on one side of the measurement context: the collateral's, since his or her report often has been considered the standard. An alternative perspective that has more heuristic promise is to view S and C reports of a subject's alcohol consumption as two independent estimates of a variable that may have no flawless measure.[21]

There are several research implications of this alternative approach to study of S and C reports of a subject's alcohol consumption. One is that factors that affect the degree of consistency between S and C reports need more research attention. For example, more research is needed on the characteristics of respondents that affect consistency. Researchers have paid some attention to subject characteristics, particularly drinking variables and demographic factors, but so far, subject factors that have been investigated primarily have yielded findings of no differences. It should be kept in mind, however, that investigations of this question, like others in this literature, are more haphazard than systematic. Therefore, different subject variables have been investigated across divergent research contexts and drinking variables. To draw firm conclusions about differences in consistency as a function of subject variables would require, for instance, a body of studies in which the same set of subject variables were examined while other factors, such as research context or drinking variables, were systematically controlled for or varied. Furthermore, some subject variables barely have been investigated; one important set of factors, subjects' cognitive functioning, has been tested only once in S and C report literature.[22] Cognitive impairment obviously can affect the accuracy of a person's self-report.

Another area that has received virtually no attention is personality characteristics of subjects that may influence consistency of S and C reports.

Although some subject variables have been investigated for their influence on S and C report consistency, C report variables have received no attention. There is no reason why the collateral side of the respondent dyad should not be studied, and this question is unanswered. The only exceptions to the lack of research attention to C report factors are the collateral's relationship to the subject and the frequency of the collateral's contact with the subject. As intuitively important as these factors might seem, they have not produced many positive findings. Some researchers found differences in consistency of S and C reports associated with frequency of subject–collateral contact in earlier studies (*see* conclusions above from data summarized in Table 1), but later empirical work showed little predictive power of the subject–collateral relationship. Relationship of collateral to subject also has not been shown to have any systematic influence on S–C report consistency. However, it is important to note that as we commented above regarding subject variables, both frequency of subject–collateral contact and the subject–collateral relationship have been studied sporadically across diverse research contexts and alcohol consumption variables. A more systematic effort of studying these factors would be more sensitive to discovering any effects on S–C report consistency that they do have.

Another important source of differences in S–C report consistency is the context in which the interview is done. Knowledge of the effects of contextual factors on the accuracy of self-reports of alcohol consumption in general has accumulated over the years.[7,21] In an interesting study that affirms current research findings on context, Sobell et al. (1991) asked subjects what interview conditions, from their perspective, would yield the most accurate information about the subject's alcohol consumption.[23] Subjects in this study were 208 problem drinkers who presented for brief intervention for their alcohol problems. As part of the pretreatment assessment, subjects reported their alcohol consumption for the year before the interview. Subjects also were asked how accurately different people they lived with or knew (e.g., spouse, relative, or friend) would describe the subject's drinking compared to his or her own self-report. This question was asked for different amounts of alcohol consumption. Furthermore, subjects were asked how accurate their self-reports would be if they had had different amounts of alcohol to drink. Finally, subjects were asked how accurate their self-reports would be if they were interviewed under different conditions. Examples are by their therapist, by a researcher they did not know, by telephone, and at home alone.

The results of this study most directly relevant to the use of C reports showed consistently that subjects felt that different collaterals would have different degrees of accuracy in their reports about a subject's drinking. For all levels of the subject's alcohol consumption except days of five to nine standard drinks, subjects felt that their spouses would provide the most accurate reports and that their employers would give the least accurate reports. For the exception of five to nine drinks, subjects thought that the person they named as primary collateral would provide the most accurate reports, and the spouse was named as next most accurate. Individuals named in the middle ranks of collateral accuracy varied with drinking level.

There are two major points to emphasize from this research. First, unlike what the literature has shown, subjects think that who the collateral is makes a difference in how consistent S and C reports are. However, as we noted, this question has not been studied systematically, and the Sobell et al. (1991) research emphasizes the need to do that. Another important point is that perceived collateral accuracy depended on another factor: the subject's drinking level. This suggests the complexity of the question of S–C report consistency and that researchers have only begun to do the kinds of studies that can help to understand that complexity. We should note that subjects' perceptions of other context variables (e.g., effects of confidentiality assurances, whether the subject is intoxicated at the time of the interview, and whether the interview is conducted in person or by telephone) were consistent with the evidence in the drinking self-report literature that is available on these factors.

Investigation of Sources of Error in Reports

Another line of research that should be pursued is sources of error in both S and C reports. Of course, such error would decrease consistency between the two report sources. Table 3 is a sample of

Table 3
Sources of Error in S and C Reports of Alcohol Consumption

Subject	Collateral
Forgetting	Forgetting
Denial	Behavior difficult to observe
Bragging	Spousal courtesy
Incentive to withhold or exaggerate information	In clinical samples, hypersensitivity about drinking and consequences

error sources in both S and C reports. The sources were collected from both empirical and review articles on S–C report consistency.

As can be seen in Table 3, information about a subject's drinking may be inaccessible for either subject or collateral recall at the time of the interview. The interviewer may or may not be able to use cueing techniques to aid retrieval. The collateral is more likely to have problems in retrieval because information about another's behavior is less salient than information about one's own. We have already discussed denial and incentive to hide or exaggerate as error sources for subjects. Less discussed are incentives to brag about alcohol consumption that may be present in certain populations, such as adolescents.

Difficulty for a collateral to observe some types of information about a subject's drinking was commonly mentioned in empirical papers on S–C report consistency. As we noted, this problem seems especially important when the respondent is asked to report on a subject's drinking in some detail or precision. Spousal courtesy refers to the reluctance of a spouse or other person close to a subject to describe a subject's drinking as particularly heavy or as causing serious problems. This source of error would most likely be present in interviews of nonclinical populations.[24] Finally, in clinical samples, there may be a tendency to magnify the amount or consequences of a subject's drinking, because of the great distress that drinking often recently has caused subjects and their loved ones who appear for treatment. This hypersensitivity would tend to result in the collateral saying that the subject drank more than he or she actually did or in exaggerating the amount of disturbance the drinking caused.

In summary, sources of error in S and C reports are extremely important research problems. Researchers have begun to talk about these error sources, but as yet, there is no systematic study of any of the sources listed in Table 3. Such research likely would result in improved S–C report consistency.

Qualitative Study of Discrepancies

A final research area that has received no attention is qualitative study of discrepancies between S and C reports.[24] A few authors have discussed discrepancies in S and C reports in order to resolve differences, as a methodological feature, but no one has conducted a qualitative study of discrepancies.[25] Such research could advance the field in at least two ways: Considering the relative lack of information on what factors influence consistency and discrepancy, qualitative research could be a source of discovery of variables worthy of quantitative investigation. In addition, it would appear that qualitative studies also would help generate hypotheses about multifactorial explanations of report consistency and discrepancy. Such theoretical models are conspicuously absent in S and C report literature.

Conclusions

We reached several conclusions on the use of C reports based on the empirical study of the question since 1986. Our conclusions relate both to the accuracy of an S report of drinking and to using a C report as a measure of a subject's drinking. First, except for the Fuller et al. (1988) study, the data consistently suggest that subjects give accurate reports about their drinking. This conclusion spans research done with various clinical and non-clinical subject populations over a range of assessment intervals (e.g., 6 mo pretreatment, or 3 mo posttreatment followup), a range of categorical and continuous measures, and the use of collaterals that had a variety of relationships (e.g., spouse, friend, or relative) with subjects. Therefore, in view of the methodological basis of the research and the different populations it includes, in general, there can be considerable confidence in the accuracy of a subject's report of his or her drinking and in the use of C reports as a measure of drinking.

This conclusion is based on two indices of consistency: percent agreement and Pearson correlations. However, the papers cited in Table 2 often included mean differences in S and C reports and discrepancy analyses when continuous measures were reported. These two indices show that typically, there were not significant differences between S and C reports in the average value of a measure. Furthermore, when discrepancies occurred there was no systematic trend for the S report to be higher or lower than the C report. These findings, of course, are consistent with the percent agreement and the correlation data.

Although in general the data showed good S–C report consistency, there were some exceptions. We referred to some of these earlier. It seems that poorer consistency is associated with more molecular, less well-defined, and harder-to-observe (for the collateral) behaviors. Also, the longer the referent recall period, the poorer the consistency that can be expected.

Consistent with earlier data, in recent research we could find no pattern of predicting S–C report agreement as a function of the frequency of contact between subjects and collaterals, the relationship of the subject to the collateral, subject demographic characteristics, or subject drinking and related factors. We should note that no study investigated all of the above variables in, say, a multiple regression model. However, the lack of power of the variables to account for variance in S–C report agreement is surprising.

A final point about the more recent data is that they not only are consistent with previous research but also extend it through inclusion of non-clinical populations. In particular, samples of the elderly and natural recoverers further increase the generalizability of the conclusion that C reports can be useful measures of drinking

The growing literature on the use of C reports shows that they are an empirically proven and sometimes essential measure in the multimethod measurement of a subject's drinking. A positive trend in the more recent literature is movement away from the view of the C report as the standard that the S report must meet. Instead, there is increasing acceptance of the perspective that the C report is another measure of a subject's alcohol consumption in addition to his or her own report of such drinking. As any measure, the C report has certain strengths and weaknesses. This reframing of what the C report is has a number of research implications, which, if pursued, will greatly increase knowledge about the use of C reports and the sensitivity of measuring a subject's alcohol consumption.

References

[1]T. Myers (1983) Corroboration of self-reported alcohol consumption: A comparison of the accounts of a group of male prisoners and those of their wives/cohabitees. *Alcohol Alcoholism* **18,** 67–74.

[2]D. Cahalan, I. H. Cisin, and H. M. Crossley (1969) *Am. Drinking Pract.* (Rutgers Center for Alcohol Studies, New Brunswick, NJ).

[3]G. A. Marlatt, R. S. Stephens, D. Kivlahan, D. J. Buef, and M. Banaji, M. (1986) Empirical evidence on the reliability and validity of self-reports of alcohol use and associated behaviors. Workshop on the Validity of Self-Report in Alcoholism Treatment Research, National Institute on Alcohol Abuse and Alcoholism, February 25, 1986, Washington, DC.

[4]R. K. Fuller, K. K. Lee, and E. Gordis (1988) Validity of self-report in alcoholism research: Results of a Veterans Administration cooperative study. *Alcoholism: Clin. Exp. Res.* **12,** 201–205.

[5]T. D. Ridley and S. T. Kordinak (1988) Reliability and validity of the Quantitative Inventory of Alcohol Disorders (QIAD) and the veracity of self-report by alcoholics. *Am. J. Drug and Alcohol Abuse* **14,** 263–292.

[6]J. A. Gladsjo, J. A. Tucker, J. L. Hawkins, and R. E. Vuchinich (1991) Adequacy of recall of drinking patterns and event occurrences associated with natural recovery from alcohol problems (Auburn University, Auburn, AL) unpublished manuscript.

[7]H. A. Skinner (1984) Assessing alcohol use by patients in treatment, in *Research Advances in Alcohol and Drug Problems,* vol. 8, R. G. Smart, H. Cappell, F. Glazer, Y. Israel, H. Kalant, R. E. Popham, W. Schmidt, and E. M. Sellers, eds. (Plenum, New York), pp. 183–207.

[8]S. A. Maisto, J. R. McKay, and G. J. Connors (1990) Self-report issues in substance abuse: State of the art and future directions. *Behav. Assess.* **12,** 117–134.

[9]L. Midanik (1982) The validity of self-reported alcohol consumption and alcohol problems: A literature review. *Brit. J. Addict.* **77,** 357–382.

[10]L. T. Midanik (1988) The validity of self-reported alcohol use: A literature review. *Brit. J. Addict.* **83,** 1019–1029.

[11]L. T. Midanik (1989) Perspectives on the validity of self-reported alcohol use. *Brit. J. Addiction* **84,** 1419–1423.

[12]T. J. O'Farrell and S. A. Maisto (1987) The utility of self-report and biological measures of alcohol consumption in alcoholism treatment outcome studies. *Advances Behav. Res. Therapy* **9,** 91–125.

[13]J. M. Polich (1982) The validity of self-reports in alcoholism research. *Addict. Behav.* **7,** 123–132.

[14]L. C. Sobell and M. B. Sobell (1986) Can we do without alcohol abusers' selfreports? *Behav. Ther.* **9**, 141–146.

[15]J. J. Bartko (1976) On various intraclass correlation reliability coefficients. *Psych. Bull.* **83**, 762–765.

[16]J. J. Bartko and W. T. Carpenter (1976) On the methods and theory of reliability. *J. Nerv. Ment. Dis.* **163**, 307–317.

[17]E. L. Spitznagel and J. E. Helzer (1985) A proposed solution to the base rate problem in the kappa statistic. *Arch. Gen. Psychiatry* **42**, 725–728.

[18]A. Anastasi (1988) *Psychological Testing*, 6th Ed. (Macmillan, New York).

[19]D. N. Nurco (1985) A discussion of validity, *Self-Report Methods of Estimating Drug Use.* B. A. Rouse, N. J. Kozel, and L. G. Richards, eds. (NIDA Research Monograph No. 57. US Government Printing Office, Washington, DC), pp. 6–11.

[20]J. A. Tucker, R. E. Vuchinich, C. V. Harris, M. G. Gavornik, and E. J. Rudd (1991) Agreement between subject and collateral verbal reports of alcohol consumption in older adults. *J. Stud. Alcohol* **52**, 148–155.

[21]T. Babor, R. Stephens, and G. A. Marlatt (1987) Verbal report methods in clinical research on alcoholism: Response bias and its minimization. *J. Stud. Alcohol* **48**, 410–424.

[22]F. Miller and A. Barasch (1985) The under-reporting of alcohol use: The role of organic mental syndromes. *Drug Alcohol Depend.* **15**, 347–351.

[23]L. C. Sobell, T. Toneatto, M. B. Sobell, G. I. Leo, and L. Johnson (1991) Going to the source: Alcohol abusers' perceptions of the accuracy of their self-reports of drinking (Addiction Research Foundation, Toronto, Ontario), unpublished manuscript.

[24]R. Room (1989) Spouse reports versus self-reports of drinking in general population surveys. Paper presented at the 15th Annual Alcohol Epidemiology Symposium, Kettil Bruun Society for Social and Epidemiological Research on Alcohol, June 11–16, 1989, Maastricht, The Netherlands.

[25]T. J. O'Farrell, H. S. G. Cutter, G. Deutch, and J. Fortgang (1984) Correspondence between one-year retrospective reports of pretreatment drinking by alcoholics and their wives. *Behav. Assess.* **6**, 263–274.

[26]E. J. Freedburg and W. E. Johnston (1980) Validity and reliability of alcoholics' self—reports of use of alcohol submitted before and after treatment. *Psychol. Rep.* **46**, 999–1005.

[27]S. B. Guze, V. B. Tuason, M. A. Stewart, and B. Pickens (1963) The drinking history: A comparison of reports by subjects and their relatives. *Quarterly J. Stud. Alcohol* **24**, 249–260.

[28]M. Hesselbrock, T. F. Babor, V. Hesselbrock, R. E. Meyer, and K. Workman (1983) Never believe an alcoholic? On the validity of self-report measures of alcohol dependence and related constructs. *Intl. J. Addict.* **18**, 593–609.

[29]K. Leonard, N. J. Dunn, and T. Jacob (1983) Drinking problems of alcoholics: Correspondence between self and spouse reports. *Addict. Behav.* **8**, 369–373.

[30]S. A. Maisto, L. C. Sobell, and M. B. Sobell (1979) Comparison of alcoholics' self-reports of drinking behavior with reports of collateral informants. *J. Consult. Clin. Psychol.* **47**, 106–112.

[31]S. A. Maisto, M. B. Sobell, and L. C. Sobell (1982) Reliability of self-reports of low ethanol consumption by problem drinkers over 18 months of follow-up. *Drug Alcohol Depend.* **9**, 273–278.

[32]T. McAuley, R. Longabaugh, and H. Gross (1978) Comparative effectiveness of self and family forms of the Michigan Alcoholism Screening Test. *J. Stud. Alcohol* **39**, 1622–1627.

[33]B. S. McCrady, T. J. Paolino, and R. Longabaugh (1978) Correspondence between reports of problem drinkers and spouses on drinking behavior and impairment. *J. Stud. Alcohol* **39**, 1252–1257.

[34]F. Miller, and A. Barasch (1985) The under-reporting of alcohol use: The role of organic mental syndromes. *Drug Alcohol Depend.* **15**, 347–351.

[35]W. R. Miller, V. L. Crawford, and C. A. Taylor (1979) Significant others as corroborative sources for problem drinkers. *Addict. Behav.* **4**, 67–70.

[36]T. Myers (1983) Corroboration of self-reported alcohol consumption: A comparison of the accounts of a group of male prisoners and those of their wives/ cohabitees. *Alcohol Alcoholism* **18**, 67–74.

[37]T. J. O'Farrell, H. S. G. Cutter, G. Dentch, and J. Fortgang (1984) Correspondence between one-year retrospective reports of pretreatment drinking by alcoholics and their wives. *Behav. Assess.* **6**, 263–274.

[38]J. B. Saunders, A. D. Wodak, A. Haines, P. R. Powell-Jackson, B. Portmann, and R. Williams (1982) Accelerated development of alcoholic cirrhosis in patients with HLA-B8. *Lancet* **19**, 1381–1384.

[39]A. W. Stacy, K. F. Widaman, R. Hays, and M. R. Dimatteo (1985) Validity of self–reports of alcohol and other drug use: A multitrait-multimethod assessment. *J. Pers. Social Psychol.* **49**, 219–232.

[40]J. S. Verinis (1983) Agreement between alcoholics and relatives when reporting follow-up status. *Intl. J. Addict.* **18**, 891–894.

[41]C. G. Watson, C. Tilleskjor, E. A. Hoodecheck–Schow, J. Pucel, and L. Jacobs (1984) Do alcoholics give valid self–reports? *J. Stud. Alcohol* **45**, 344–348.

[42]L. J. Davis and R. M. Morse (1987) Patient-spouse agreement on drinking behaviors of alcoholics. *Mayo Clinic Proc.* **62**, 689–694.

[43]G. J. Loethen and K. A. Khavari (1990) Comparison of the Self-Administered Alcoholism Screening Test (SAAST) and the Khavari Alcohol Test (KAT): Results from an alcoholic population and their collaterals. *Alcoholism: Clin. Exp. Res.* **14**, 756–760.

[44]L. C. Sobell, M. B. Sobell, and T. Toneatto (in press) Recovery from alcohol problems without treatment, *Self-Control and Addictive Behaviors*. N. Heather, W. R. Miller, and J. Greeley eds. (Pergamon, New York).

[45]G. Wolber, W. F. Carne, and R. Alexander (1990) The validity of self-reported abstinence and quality sobriety following chemical dependency treatment. *Intl. J. Addict.* **25**, 495–513.

BIOCHEMICAL MEASURES
OF ALCOHOL CONSUMPTION

An Overview of Current and Emerging Markers of Alcoholism

Alan S. Rosman and Charles S. Lieber

Introduction

Types of Biological Markers of Alcoholism and Their Applications

The main purpose of biological markers of alcohol consumption is to provide valuable tools in the diagnosis and treatment of alcoholism. One can categorize different types of biological markers of alcoholism based on their characteristics and clinical applications as trait markers, state markers, and markers of organ damage.

Trait Markers

Trait markers, also known as markers of predisposition to alcoholism, should determine which individuals are at risk of becoming alcoholics. These markers would provide an important research tool in evaluating the genetic and environmental factors that may predispose to alcoholism.

State Markers

State markers should reflect an individual's consumption of alcohol and can be further subdivided into markers of chronic consumption (screening markers) and markers of acute consumption (relapse markers).

From: *Measuring Alcohol Consumption*
Eds.: R. Litten and J. Allen ©1992 The Humana Press Inc.

Markers of Chronic Consumption. Markers of chronic consumption would be of value in the screening patients who are drinking at levels that could result in long-term behavioral or medical problems. These markers would be useful in the early identification of alcoholism as many alcoholic patients do not admit their drinking problems to their physicians.[1] Furthermore, physicians are only able to identify a small percentage of alcoholics attending a general medical clinic.[2] A positive screening marker of excessive alcohol consumption could alert the physician to further explore a patient's drinking history and initiate early intervention if indicated. In addition, a marker of excessive alcohol consumption would have important diagnostic uses. Many patients suffering from a disease for which excessive alcohol consumption is the suspected etiologic agent, e.g., chronic pancreatitis, cirrhosis, or cardiomyopathy, fail to give an accurate history of their alcohol consumption. Many of these patients are then subjected to an exhaustive diagnostic workup to rule out other etiologic factors. A reliable biological marker of excessive alcohol consumption would expedite the workup and perhaps avoid unnecessary invasive procedures.

In addition, a reliable biological marker would be a valuable tool for epidemiologic studies investigating incidences of alcoholism in various communities. Potential forensic applications include screening individuals arrested for driving while intoxicated for chronic alcoholic consumption.[3,4] Drivers who have markers suggesting a chronic problem could be referred to an appropriate treatment program. These markers could also screen workers involved in areas of public safety for excessive alcohol consumption.

Markers of Acute Consumption. A marker of acute consumption should be able to detect early relapses in alcoholic patients undergoing treatment. Such a marker should be sensitive to low levels of drinking since the goal of most treatment programs is complete abstinence. Several studies have demonstrated the lack of reliability of a patient's self-report of alcohol consumption for diagnosing relapse.[5-7] In a Veteran's Administration Cooperative study evaluating the efficacy of disulfiram, approximately 35% of patients claiming to be abstinent were indeed drinking during the study period as confirmed by collateral history and/or laboratory tests.[7] Thus, a marker of recent drinking could be used to monitor sobriety in alcoholic patients in outpatient treatment. A positive test would alert the physician of an early relapse and could then prompt treatment at an early stage.

Breath alcohol testing is currently used in many treatment programs for monitoring sobriety. Because the rate of ethanol metabolism is about 10 g/h, breath alcohol testing will only identify recent intake, usually within a few hours after the last drink.[8] An ideal marker of relapse should remain elevated for at least a few days after the patient's last drink.

Finally, a biological marker of relapse would be valuable in clinical research. Currently, clinical trials evaluating the efficacy of various treatment modalities usually rely on the patient's self-report of alcohol consumption as the therapeutic endpoint. A biological marker of relapse would provide a more objective assessment of sobriety.

Markers of Organ Damage
from Alcohol Consumption

Excessive alcohol consumption can result in chronic injury to a variety of organs. The identification of patients with early organ injury, e.g., precirrhotic lesions, could result in intensive abstinence therapy and may prevent the morbidity associated with end-stage, irreversible organ damage.

Diagnostic Characteristics
of Laboratory Tests

Physicians primarily use laboratory tests when clinical evaluation results in diagnostic uncertainty. The usefulness of a laboratory marker of alcohol consumption lies in its ability to accurately discriminate between patients who are actively drinking and nondrinking individuals. The diagnostic characteristics used to assess the discriminating ability of laboratory tests include sensitivity and specificity.[9-11] In marker studies, sensitivity is the proportion of patients who are actively drinking above a given threshold and have a positive (abnormal) test. Specificity is the proportion of patients who are drinking below that threshold and have a negative (normal) test.

Although the threshold level of "normal alcohol consumption" is often arbitrary, it can reflect the level of drinking that can increase the risk of organ damage.

In utilizing laboratory tests for diagnostic purposes, physicians tend to use the predictive values of a test. The positive predictive value is the probability that the patient is a heavy consumer of alcohol when a marker is positive. The negative predictive value is the probability that the patient is not a heavy consumer of alcohol when a marker is negative. By applying Bayes' theorem, the positive and negative predictive values can be derived from the sensitivity, specificity, and pretest probability, which is equivalent to the prevalence of heavy alcohol consumption when a marker is used to screen a population. When a marker is used in the diagnostic evaluation of a particular patient, the pretest probability is the physician's estimation of the likelihood of the patient being a heavy alcohol consumer prior to laboratory testing. In Fig. 1, the predictive value of a hypothetical marker of excessive alcohol consumption with a sensitivity and specificity of 90% is plotted as a function of the prevalence of excessive alcohol consumption. At a prevalence of 10%, the predictive value of the positive test is only 50%, which may not be high enough for many clinical situations. However, if clinical evaluation suggested that the patient had a 50% likelihood of excessive alcohol consumption prior to laboratory testing, the predictive value of a positive test would be 90%. In this setting, the test would be useful in corroborating a clinical suspicion of excessive alcohol consumption.

For maximal clinical application, the ideal marker should have a high enough sensitivity and specificity to be a useful screening tool. A screening marker of excessive alcohol consumption should also be able to discriminate between social drinking and excessive alcohol consumption. In contrast, a marker of relapse should be sensitive to even moderate drinking, since the goal of most treatment programs is complete abstinence. The possible causes of false-positive tests should be readily identified and easy to exclude in light of the social stigma associated with the diagnosis of alcoholism. Thus, an ideal marker should not be elevated by either nonalcoholic liver disease or nutritional status. Also, a marker should be relatively noninvasive, e.g., a blood, urine, or breath test.

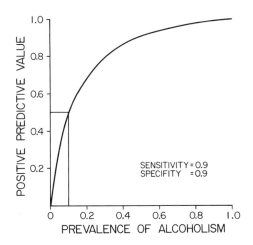

Fig. 1. The positive predictive value of a positive test for a theoretical marker is plotted as a function of the prevalence of alcoholism using Bayes' theorem. The theoretical marker is assumed to have a sensitivity and specificity of 90%. If the marker was used to screen a population with a 10% prevalence of alcoholism, the predictive value of a positive test would be only 50%, from "Biological Markers of Alcoholism" (Rosman and Lieber), in *Medical and Nutritional Complications of Alcoholism,* edited by C. S. Lieber (Plenum, in press).

Biological Markers
of a Predisposition to Alcoholism

Research during the last two decades has suggested a genetic predisposition to alcoholism. Two forms of alcoholism have been identified. Type 1 is usually adult onset, has fewer social problems, occurs in both sexes, and is environmentally determined, whereas Type 2 is usually early onset, is associated with antisocial behavior, occurs mainly in males, and has a strong genetic predisposition.[12] In light of these findings, a biological marker of a predisposition to alcoholism would provide further understanding of the interaction of genetic and environmental factors in the pathogenesis of alcoholism. Potential candidates for a predisposition marker include platelet enzyme activities, lymphocyte markers, and genetic mapping.

Autopsy studies have suggested that alcoholics may have decreased central nervous system (CNS) monoamine oxidase (MAO) activity.[13,14] Platelet MAO may reflect the levels of MAO in the CNS. Studies have reported that platelet MAO activity was decreased in Type 2 alcoholics but not in Type 1.[15,16] Studies have also reported that platelet MAO may remain depressed in alcoholics despite several years of abstinence.[17,18] Other platelet enzymes, e.g., adenylate cyclase activity, may also be altered in alcoholics.[19]

It has been postulated that alcoholism is associated with abnormalities in the receptor-mediated transmembrane signalling system in cells of the CNS.[20] Lymphocytes also have adenosine receptors and their activation can result in increased cyclic adenosine monophosphate (cAMP) levels.[21] Studies have reported that lymphocytes from alcoholics have altered basal and adenosine receptor-stimulated cAMP levels as compared to controls.[22,23] The use of DNA probes and endonuclease mapping could potentially identify restriction fragment length polymorphisms associated with alcoholism.[24] Using this approach, Blum et al. (1990) identified an allelic form of the dopamine D_2 receptor gene associated with alcoholism.[25] However, another study failed to confirm this association.[26] The conflicting results may be related to the different methods used in patient selection.

Limitations of Current Markers of Alcohol Consumption

Currently used markers of excessive alcohol consumption include γ-glutamyl transpeptidase (GGT), mean corpuscular volume (MCV), the transaminases, and high-density lipoprotein-cholesterol (HDL-C). Alpha amino-*n*-butyric acid (AANB) has also been used as a marker. These tests are occasionally useful in confirming a clinical suspicion of alcoholism, but they have several limitations.

γ-Glutamyl Transpeptidase

Chronic ethanol administration has been shown to increase serum GGT in experimental animals and in alcoholic patients.[27–30] The most common mechanisms for the increase in serum GGT by ethanol include induction of hepatic GGT and increased lability of hepatic plasma membranes.[27, 29-31] However, other possibilities such as increased GGT synthesis secondary to cholestasis and damage to hepatocytes may be seen in advanced alcoholic liver damage.[32,33]

Several studies have investigated the sensitivity of GGT as a marker of alcoholism. After pooling most of these studies, we determined that the sensitivity of GGT for hospitalized alcoholics is 62% as compared to 43% for ambulatory alcoholics.[34] The increased sensitivity in hospitalized patients may be secondary to their more advanced state of alcoholism.[34] Serum GGT may also be elevated by nonalcoholic conditions including microsomal inducing agents (e.g., anticonvulsants), nonalcoholic liver disease, and biliary tract disease.[35] Thus, GGT is not an ideal screening marker because of a lack of sensitivity and moderate specificity. However, it is useful in confirming a clinical suspicion of alcoholism and may be of value in monitoring alcoholics for relapse although it is not ideal for that purpose.[36-38]

Mean Corpuscular Volume

Chronic alcohol consumption can occasionally elevate MCV. The most common mechanism may be related to a direct toxic effect of ethanol.[39-41] However, other mechanisms—including folic acid deficiency and macrocytosis secondary to liver disease—may also occur. After pooling several studies, we determined that the sensitivity of the MCV for detecting alcoholism is only 35–40%.[34] An elevated MCV may also occur in nonalcoholic conditions such as nutritional deficiencies (B_{12}, folate), nonalcoholic liver disease, reticulocytosis, and hematological malignancies.[42] As these conditions are not relatively common in an ambulatory setting, an elevated MCV is usually specific for alcoholism. Despite its specificity, MCV is not an ideal marker for alcoholism because of its low sensitivity, and MCV is usually not helpful in monitoring alcoholics for relapse because of its prolonged half-life.[36]

Serum Transaminases

Two commonly performed tests are the serum transaminases: aspartate aminotransferase (AST) and alanine aminotransferase (ALT). ALT and AST are abundant in the liver, and AST is also present in muscle and myocardial cells.[43] Chronic alcohol administration can occasionally result in elevations of the transaminases either by increasing the cell membrane permeability or by cell necrosis.[44] Pooling several studies, we determined that the sensitivity of serum AST as a marker of alcoholism is only 35%.[34] Serum ALT is even less sensitive than AST. Furthermore, both transaminases may be elevated in nonalcoholic liver disease, and AST may be elevated in muscular disorders and in myocardial injury.[43]

High-Density Lipoprotein-Cholesterol

Serum HDL-C can be elevated by both moderate ethanol consumption (20–40 g/d) and high ethanol consumption (more than 70 g/d).[45,46] Mechanisms for increased serum high density lipoprotein (HDL) secondary to alcohol consumption have been attributed to the following: induction of microsomal enzymes with increased synthesis of apolipoproteins A-I and A-II and increased plasma lipoprotein lipase activity resulting in an accelerated turnover of very low density lipoprotein (VLDL) and an increased production of HDL_2.[47–52] In addition, severe liver disease can attenuate the effects of alcohol consumption on HDL-C.[53]

The complex effects of alcohol on HDL metabolism and the high variability of serum HDL-C in the normal population limit the utility of serum HDL-C as a marker of alcoholism.[54] After pooling several studies, we estimated the sensitivity of serum HDL-C to be only 30%.[34] Furthermore, serum HDL-C can be elevated by moderate drinking, exercise, diet, genetic factors, and medications.[55,56] Table 1 summarizes the characteristics of serum HDL-C and other currently used markers.

α-Amino-N-Butyric Acid

Alpha amino-n-butyric acid (AANB) is a nonessential amino acid derived from the catabolism of methionine, serine, and threonine.[57] Studies have demonstrated that ethanol administration elevated plasma AANB in both experimental animals and in alcoholic patients.[58,59] Because AANB can be affected by a variety of nutritional and metabolic factors, it has limited utility as a screening marker of alcoholism. However, Shaw et al. (1979) reported that sequential plasma

Table 1
Characteristics of Current Laboratory Markers of Alcohol Consumption

Marker	Estimated sensitivity	Half-life decay	Causes of false-positive
GGT	50%	26 d (variable)	Nonalcoholic liver disease, biliary tract disease, anti-convulsants,anticoagulants, hyperlipidemia, hyperthyroidism, obesity
MCV	35–40	Mo (variable)	B_{12} deficiency, folate deficiency hypothyroidism, malignancies, nonalcoholic disease
AST	35	Wk (variable)	Nonalcoholic liver disease, muscle disorders, myocardial infraction
ALT	30–35	Wk (variable)	Nonalcoholic liver disease
HDL-C	30	5–10 d	Genetic factors, diet, exercise, medications, social drinking, gender

AANB measurements could be useful in monitoring alcoholic patients for relapses, particularly when compared to a baseline AANB obtained after 3 wk of inpatient treatment (Fig. 2).[36]

Limitations
of Using Combinations of Markers

Currently, no single marker has been shown to have sufficient diagnostic accuracy to be useful for the screening of ambulatory patients for alcoholism.[35] Several investigators have combined markers to improve diagnostic accuracy. In a qualitative combination, combining the markers can increase either sensitivity or specificity. As shown in Table 2, requiring that only one of the markers be positive for a diagnosis of alcoholism, increases the sensitivity but decreases specificity; requiring that all of the tests be positive for a diagnosis of alcoholism increases specificity but decreases sensitivity.

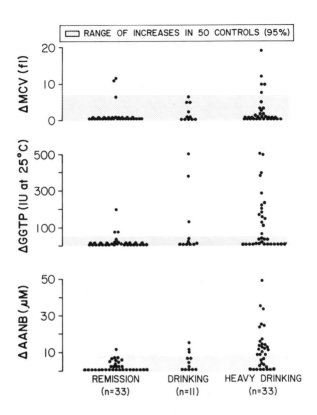

Fig. 2. Increases in MCV, GGT, and AANB above the discharge value in alcoholics who relapsed as compared to alcoholics in remission. A majority of alcoholics who relapsed had an increase in either GGT or AANB,[36] from "Detection of alcoholism relapse: comparative diagnostic value of MCV, GGTP, and AANB," in *Alcoholism: Clinical and Experimental Research,* Williams & Wilkins, 1979, **3**, 299.

Table 2
Qualitative Combination of Markers[156]

Marker	Sensitivity	Specificity
GGT	50%	78%
MCV	32	94
GGT or MCV	63	74
GGT and MCV	17	98

An alternative approach is using discriminant analysis to mathematically combine markers to generate a formula that can be used for diagnostic classification. Table 3 summarizes the findings of studies using linear discriminant analysis in the diagnosis of alcoholism. It is apparent from Table 3 that the studies selected different markers for their discriminant functions because of differences in the study populations. Furthermore, none of these studies reported a sensitivity greater than 90%. It is uncertain whether quadratic discriminant analysis offers any advantage over linear discriminant analysis.[60,61] Discriminant analysis using currently available markers does not appear accurate enough to be used as a screening test.

Special Laboratory Markers
of Alcohol Consumption

A number of special laboratory markers of alcoholism have been developed that can have increased diagnostic accuracy as compared to currently available markers. The diagnostic features of these tests are summarized below.

Carbohydrate-Deficient Transferrin

Recent studies have suggested that ethanol may have important effects on transferrin metabolism. When secreted into the blood, transferrin normally possesses a variety of carbohydrate moieties including sialic acid, galactose, *N*-acetylglucosamine, and mannose.[62] When analyzed by isoelectric focusing (IEF), the main component of transferrin has an isoelectric point (pI) of 5.4, but alterations in the sialic content of transferrin will the change the pI.[63] Stibler and Borg (1986) reported that chronic alcohol consumption will reduce the number of carbohydrate moieties attached to serum transferrin, producing carbohydrate-deficient transferrin (CDT).[63] This variant form of transferrin has an increased pI (5.7–5.9) because of decreased sialic acid content and accounted for 5–10% of transferrin in the serum of alcoholics.[64–69] The mechanism for the alteration in transferrin is unclear but has been attributed to reduced glycoprotein glycosyltransferase activity associated with alcohol consumption.[70]

Table 3
Studies Using Linear Discriminant Analysis to Identify Alcoholics

Reference	Population studied	Laboratory values selected for discriminant function	Sensitivity %	Specificity %
157*	Alcoholics, gastroenterology clinic patients and patients with nonalcoholic liver disease	MCV, GGT, alkaline phosphatase	87	94
158	Alcoholic and nonalcoholic Swedish military recruits	MCV, GGT, ALT, AST, uric acid apolipoprotein A1, ferritin	54	94
159	Alcoholic and nonalcoholic patients participating in the British Regional Heart Study	GGT, HDL-C, uric acid, lead, mean corpuscular hemoglobin	41	98
160	Young–adult alcoholics and nonalcoholic college students visiting Student Health Clinic	BUN, potassium, MCV	89	92
61*	Hospitalized alcoholics and nonalcoholic and nonalcoholic psychiatric inpatients	BUN, AST, calcium, phosphorus, albumin	59	72

*Sensitivity and specificity were determined by cross-validation.

Different laboratory tests have been developed for measuring CDT. A quantitative method using IEF followed by immunofixation (IF) with antitransferrin antibodies performed directly on the focused gel has been described in several studies.[69,71,72] After conventional staining, densitometry is then used to quantify the percentage of the abnormal transferrin (pI of 5.7) relative to the total amount of immunofixed transferrin (CDT–transferrin ratio). However, this method of IEF is not well suited for routine laboratory purposes and usually does not measure the absolute amount of CDT in the serum.[73] Thus, methods involving anion chromatography have been developed to separate CDT from the main component of transferrin followed by radioimmunoassay (RIA) using transferrin antibodies.[38,73–76] For potential laboratory application, a kit consisting of disposable microcolumns and reagents for the RIA has been developed (Pharmacia Diagnostics AB, Uppsala, Sweden); however, these kits are not yet commercially available. Recently, a sensitive and practical method employing IEF followed by Western blotting (WB) has been developed.[77] In this technique, the focused proteins are transferred electrophoretically from the gel to a nitrocellulose membrane. Immunochemical staining of the nitrocellulose membrane is then performed using goat antihuman transferrin IgG and biotin-conjugated rabbit antigoat IgG. The absolute amount of CDT can be measured using computerized scanning densitometry of the immunoreaction intensity. This technique allows complete resolution of the CDT from the normal transferrin components and is also quite sensitive, since nitrocellulose membranes are usually more accessible to antibodies than the gels used for IEF.[78] This method also avoids the use of radioactive reagents, which have a short shelf life and require specialized handling. Thus, this technique is potentially applicable for widespread laboratory use.

A variety of studies have evaluated the sensitivities of the CDT–transferrin ratio (Table 4) and the absolute value of serum CDT (Table 5) in detecting excessive alcohol consumption. Most studies reported that the absolute value of CDT and the CDT–transferrin ratio had sensitivities of at least 80% in evaluating alcoholics admitted for detoxification. Furthermore, these tests were far more sensitive than other conventional markers of alcoholism.[72–76,79] However, in a study of ambulatory alcoholic (defined by an intake in excess of 80 g/d of

Table 4
Studies Evaluating CDT:Transferrin Ratio as a Marker of Alcohol Consumption

Reference	Method	Increased ratio %	Selection of alcoholics
71	IEF/IF	81	Patients referred for detoxification
74	Chromatography/RIA	90	Patients referrred for detoxification
75	Chromatography/RIA	56	Patients admitted for detoxification
82	IEF/IF	81	Patients admitted for detoxification
80	IEF/IF	45	Outpatients screened for alcoholism
76	Chromatography/RIA	81	Patients admitted for detoxification

Note: IEF = Isoelectric focusing; IF immunofixation; and RIA = Radioimmune assay

Table 5
Studies Evaluating CDT as a Marker of Alcohol Consumption

Reference	Method	Increased CDT %	Selection of patients
73	Chromatography/RIA	91	Patients admitted for detoxification
75	Chromatography/RIA	81	Patients admitted for detoxification
72	IEF/IF (Qualitative)	86	Patients admitted for detoxification
82	Chromatography/RIA	76	Patients admitted for detoxification
76	Chromatography/RIA	85	Patients admitted for detoxification
79	IEF/WB	80	Patients admitted for detoxification

Note: RIA = Radioimmune assay; IEF = Isoelectric focusing; IF = immunofixation; and WB = Western Blotting.

ethanol), the sensitivity of the CDT–transferrin ratio was only 45%.[80] Further studies are needed to evaluate the use of CDT and/or CDT–transferrin ratio in ambulatory alcoholics.

Elevations of serum CDT appear to be specific for excessive alcohol consumption. Serum CDT is elevated neither by moderate alcohol consumption (approximately 40 g/d) nor by various medications that can elevate other conventional markers of alcoholism, e.g., anticonvulsants, antipsychotics, and anticoagulants. Stibler et al. (1986) reported that CDT had a specificity of 98% when applied to a population of patients with a variety of nonalcoholic medical disorders.[73] CDT may be of value in discriminating recently drinking alcoholics from patients with nonalcoholic liver disease. Studies have reported that neither the absolute value of CDT nor the CDT–transferrin ratio were elevated in non-alcoholic patients with liver disease.[72,74,76,81,82] However, Behrens et al. (1988) reported that 38% of nondrinking patients with primary biliary cirrhosis had elevated CDT levels (Fig. 3).[75] Preliminary studies at our institution have suggested that abstinent alcoholics with cirrhosis may have elevated CDT–transferrin ratios because of depressed serum transferrin levels.[77] Stibler and colleagues (1986) reported a significant correlation between CDT values and the daily alcohol consumption in a population that included "normal consumers" and alcoholics ($r = 0.64$).[73] However, the correlation was only 0.22 in their subgroup of alcoholic patients. Behrens et al. (1988) reported no significant correlation in their population of alcoholic patients.[75]

In most alcoholic patients, CDT levels decreased progressively with cessation of drinking with a biological half-life of 16–17 d (Fig. 4).[73,75] However, some alcoholics may have CDT levels that fluctuate during the abstinent period rather than progressively decline.[75] Further studies are therefore needed to determine the utility of CDT in monitoring relapse, though the results thus far obtained are encouraging. In conclusion, initial studies suggest that CDT may be a useful screening marker of alcoholism because it has a higher sensitivity and specificity than other conventional markers of alcoholism. The development of an IEF–WB method may now allow widespread application of this test.

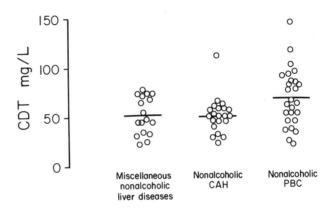

Fig. 3. CDT levels in patients with nonalcoholic liver disease. In this study performed at the Bronx Veterans and Mount Sinai Medical Centers in New York, most patients with nonalcoholic liver disease had normal CDT levels. However, 38% of patients with primary biliary cirrhosis had elevated CDT values,[75] from "Carbohydrate-deficient transferrin, a marker for chronic alcohol consumption in different ethnic populations, in *Alcoholism: Clinical and Experimental Research,* Williams & Wilkins, 1988, **12,** 429.

Red Blood Cell-
and Hemoglobin-Associated Acetaldehyde

Plasma acetaldehyde is not a reliable marker of ethanol consumption because acetaldehyde is metabolized within a few hours after ethanol consumption.[84,85] Using improved methods for determining acetaldehyde, studies have reported that a significant amount of acetaldehyde may be reversibly bound to erythrocytes.[86,87] Peterson and Polizzi (1987) utilized a fluorigenic high-pressure liquid chromatography (HPLC) assay based on the reaction of acetaldehyde with 1,3-cyclohexanedione to measure hemoglobin-associated acetaldehyde;[86] in contrast, Baraona and coworkers (1987) utilized the head-space gas chromatography method after semicarbazide extraction to measure red blood cell (RBC)-associated acetaldehyde.[87] Both hemoglobin-associated acetaldehyde and RBC-associated acetaldehyde may persist for several days after consuming ethanol.[88,89] Further clinical studies are needed to determine whether these tests will be of value as screening and/or relapse markers. A recent study reported that abstinent

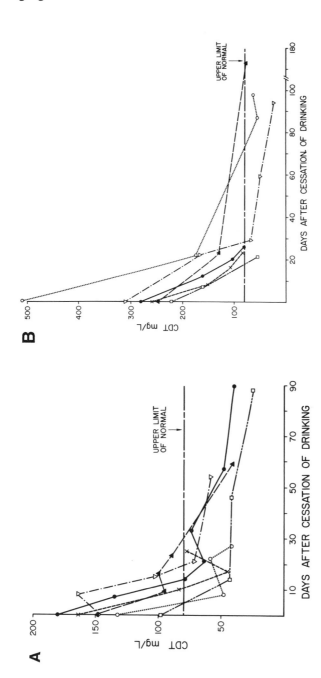

Fig. 4. Sequential CDT levels in alcoholics with an initial CDT value below 200 mg/L, Panel A, or greater than 200 mg/L, Panel B,[38] from "Changes in carbohydrate-deficient transferrin levels after alcohol withdrawal," in *Alcoholism: Clinical and Experimental Research*, Williams & Wilkins, 1988, **12**, 540.

alcoholic patients with cirrhosis and nonalcoholic patients with liver disease may have elevated RBC acetaldehyde, limiting the application of this test as a marker of chronic consumption.[90] However, its usefulness as a marker of acute consumption (to be applied for the detection of relapse) should be explored.

Red Blood Cell Cysteine

Hernandez-Munoz et al. (1989) reported that RBCs from alcoholic patients have an increased binding capacity for acetaldehyde as compared to controls.[89] Further investigation suggested that this increased binding capacity was the result of an increased capacity of the nonprotein components of RBCs and was inhibited by amino and thiol blockers. It was postulated that the increased binding capacity associated with alcoholism was due to increased cysteine, which could form an acetaldehyde adduct (AA) 2-methylthiazolidine. Using a spectrophotometric method, Hernandez-Nunoz et al. reported that RBC cysteine was elevated in most alcoholic patients. As shown in Fig. 5, elevations in RBC cysteine may persist for several weeks following alcohol withdrawal. A preliminary study reported that the sensitivity of RBC cysteine as a marker of alcoholism was 81%, superior to conventional markers.[79] However, RBC cysteine may also be elevated in non-drinking patients with liver disease.[91]

Ethanol and Acetaldehyde Adducts

Both ethanol and acetaldehyde may form stable chemical adducts with a variety of biological substrates. Phospholipid ethanol adducts have been identified in various tissues of laboratory animals that were administered ethanol.[92,93] Acetaldehyde may react with a variety of nucleophilic groups in amino acids and proteins.[94,95] In many cases, the first reaction may result in a Schiff base; because Schiff bases are chemically unstable, a second reaction, e.g., reduction or rearrangement, is required to generate a stable adduct.[94] The characterization of biologically stable acetaldehyde adducts in vivo may possibly serve as a marker of chronic ethanol consumption analogous to the use of glycosylated hemoglobin as a marker of glucose control in diabetics. A variety of biological compounds may form adducts with acetaldehyde, including amines, plasma proteins, hemoglobin,

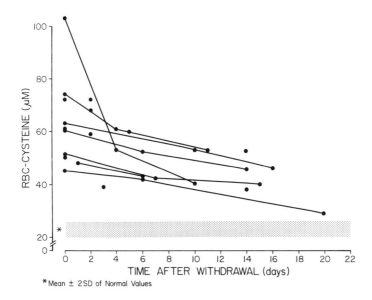

Fig. 5. Evolution of RBC cysteine in alcoholics following alcohol withdrawal. Elevations in RBC cysteine may persist for several days or weeks after cessation of drinking. Normal values ±2 SD are indicated by the shaded area,[89] from "Characterization of the increased binding of acetaldehyde to red blood cells in alcoholics," in *Alcoholism: Clinical and Experimental Research,* Williams & Wilkins, 1989, **13,** 657.

and intracellular proteins such as microsomal P450IIE1 and collagen.[96,97] Furthermore, some of these acetaldehyde adducts may act as neoantigens, triggering antibody formation *(vide infra).*

Acetaldehyde Adducts
with Amino Acids and Amines

As previously discussed, a study by Hernandez-Munoz et al. (1989) suggested that acetaldehyde may form an adduct with cysteine to generate 2-methylthiazolidine.[89] The development of a sensitive and practical assay of RBC 2-methylthiazolidine could be of clinical utility. Acetaldehyde may also form condensation products with catecholamines and indoleamines.[98–103] However, the use of these conjugates as markers may be limited because of the degree of overlap between alcoholics and controls.[98,101,104]

Acetaldehyde Protein Adducts

In vitro studies have demonstrated that acetaldehyde can form stable adducts with proteins; this was first shown for hepatic microsomal proteins and for hemoglobin.[105,106] Other proteins include albumin and low-density lipoproteins.[107–110] Lumeng and Durant (1985) further demonstrated that acetaldehyde can form stable adducts with human serum albumin at physiologic concentrations of acetaldehyde, i.e., similar to concentrations found in alcoholics during acute alcohol consumption.[108] Studies have demonstrated acetaldehyde-liver protein adducts in ethanol-fed rats.[96,111] Behrens et al. (1988) further characterized one of the liver AAs as a derivative of P450IIE1.[96] In a recent study, a sensitive, two-site enzyme-linked immunoabsorbent assay (ELISA) was used to detect acetaldehyde-serum protein adducts in alcoholic patients.[112] Seventy-one percent of recently drinking alcoholics had detectable serum protein-AAs, which persisted for up to 2 wk in some patients. Further characterization of these adduct suggested that the protein components were distinct from albumin.[112] These studies suggest that immunochemical assays of protein-AAs may lead to useful markers. A potential problem, however, may result from endogenous acetaldehyde formation, especially in view of the fact that this is increased in nonalcoholic liver disease.[90,113]

Acetaldehyde Hemoglobin Adducts

Several studies have reported that hemoglobin adducts may form in vitro when using chemical reducing agents.[106,108,114] However, the quantification of in vivo acetaldehyde hemoglobin adducts has been elusive. Several in vitro studies have reported that incubation of hemoglobin with acetaldehyde will increase the fast-migrating fraction when analyzed by cation-exchange chromatography.[106,115–117] However, the application of column chromatography for in vivo identification of hemoglobin adducts has yielded conflicting results. Several studies have reported an increase in the fast-migrating hemoglobin fractions in alcoholic patients, but other studies failed to detect any difference from controls.[106,116–122] Other chemical techniques such as IEF and HPLC analysis of tryptic digests have been inconsistent in identifying hemoglobin adducts in vivo.[120,123] An ELISA method was recently developed that detected hemoglobin adducts in volunteers following acute consumption of a large dose of ethanol.[124] Further studies are needed to determine whether immunochemical methods of hemoglobin adducts will result in an accurate marker of alcohol consumption.

Antibodies Against Acetaldehyde Adducts

Israel and colleagues (1986) reported that AAs may act as neoantigens, generating antibodies specific for acetaldehyde epitopes in laboratory mice.[125] Clinical studies have also reported elevated titers of antibodies against AAs in alcoholic patients.[126,127] However, as shown in Fig. 6, patients with nonalcoholic liver disease also had elevated titers, limiting the clinical application of this test as a marker of alcoholism.[128]

Other Emerging Markers

Mitochondrial AST

Total aminotransferase of hepatic origin consists of two isoenzymes: mitochondrial aminotransferase (mAST) and cytosolic aminotransferase.[129] Because alcohol consumption results in selective injury to mitochondria, mAST may be preferentially released into the serum.[130] Nalpas and coworkers (1986) reported that 84% of alcoholic patients had elevated serum mAST levels.[131] As patients with nonalcoholic liver disease also had elevated serum mAST, Nalpas et al. improved specificity by using the ratio of mAST to total AST.[131]

β–Hexosaminidase

Ethanol consumption may result in increased levels of serum β-hexosaminidase, a hepatic lysosomal enzyme.[132] Studies have reported that serum β-hexosaminidase was elevated in at least 85% of alcoholics.[133,134] However, this enzyme was also elevated in nonalcoholic liver disease and in pregnancy.[135–138]

Urinary Dolichols

Dolichols are long-chain polyisoprenoid alcohols that function as carrier lipids in the biosynthesis of glycoproteins.[139] Studies have reported that alcohol consumption increases urinary dolichol levels.[140,141] However, the clinical utility of dolichols as a marker may be limited because of its short biological half-life and lack of specificity for alcohol consumption.[142,143]

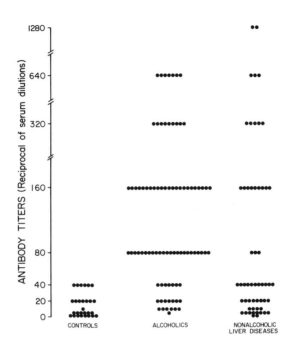

Fig. 6. Serum antibody titers against AAs in healthy controls, alcoholics, and patients with nonalcoholic liver disease,[128] from "The role of alcoholism and liver disease in the appearance of serum antibodies against acetaldehyde adducts," *Hepatology*, 1988, **8**, 569–574.

Markers of Organ Damage

Long-term alcohol consumption can result in chronic damage to a variety of organs, e.g., liver, brain, pancreas, and heart, producing significant morbidity and mortality. Abstinence-promoting therapy is usually more effective at early stages of organ damage. For example, long-term abstinence is much more likely to normalize life expectancy in patients with alcoholic cirrhosis without evidence of hepatic decompensation than in patients with advanced cirrhosis.[144] Thus, markers of alcohol-related liver injury would be of value in identifying early liver injury prior to the development of portal hypertension or hepatic insufficiency. Blood markers of alcoholic liver injury have been recently reviewed by Lieber (1988) and are summarized below.[145]

Markers of Hepatic Fibrosis

Studies have identified perivenular fibrosis (PVF) as a precursor lesion of alcoholic cirrhosis in baboons and in humans.[146,147] In the clinical study shown in Fig. 7, 83% of alcoholics with PVF developed further histological progression over a 2-yr period with continued drinking. In contrast, only 16% of alcoholics with simple steatosis developed histological progression with continued drinking. These studies suggest that alcoholic patients with PVF have a high risk of further histological progression, particularly if they continue to drink excessively.

Currently, the only reliable method for detecting PVF and early cirrhosis is liver biopsy.[148] However, this is not practical for widespread screening of alcoholic patients. The development of a noninvasive marker of PVF and early cirrhosis would identify patients who are at risk of developing further histological progression and/or hepatic decompensation with continued drinking. Recently, Lieber et al. (1990) reported that polyunsaturated phospholipid prevented further histological progression in baboons with PVF despite continued ethanol administration.[149] Thus, a noninvasive marker of PVF would be of value in identifying patients for a clinical trial evaluating this therapeutic agent.

Serum liver enzymes are not highly accurate in identifying PVF and early cirrhosis.[150,151] Because of this, attempts were made to develop serological markers related to substances involved in collagen metabolism. An RIA of the aminoterminal propeptides of type III procollagen (P-III-P) was initially evaluated, but failed to differentiate patients with fatty liver from those with early fibrosis.[152] A modified P-III-P assay using Fab (antigen binding) fragments of the antibody (Fab-P-III-P) has been demonstrated to detect not only the intact P-III-P propeptide but also peptide fragments, the major components of P-III-P-peptides.[153] This test was of value in discriminating alcoholics with PVF and cirrhosis from those with steatosis (Fig. 8).[151,153] Values of Fab-P-III-P above 45 ng/mL detected 55% of the patients with PVF, 62% of cases with septal fibrosis, and 90% of cirrhotics. However, alcohol withdrawal may be associated with increasing Fab-P-III-P values.[153] Thus, it is recommended to obtain this test during the first week of alcohol withdrawal for optimal clinical utility.

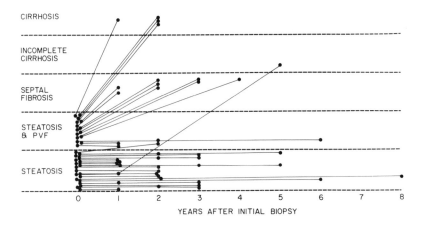

Fig. 7. Progression of fibrosis in alcoholics without hepatitis followed up to 8 yr after initial liver biopsy. Presence of PVF on the initial biopsy suggested a greater risk of further histological progression,[147] from "Perivenular fibrosis as precursor lesion of cirrhosis," *JAMA*, 1985, **254**, 627–630.

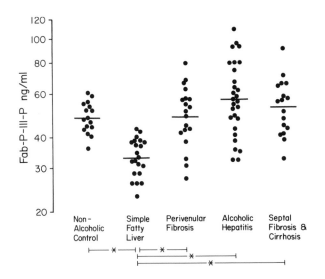

Fig. 8. Serum Fab-P-III-P in patients with alcoholic liver disease. Fab-P-III-P distinguishes a majority of actively drinking alcoholics with fibrosis from those with simple fatty liver,[153] from "Serum Procollagen Type III N-terminal peptides and laminin peptide in alcoholic liver disease," in *Alcoholism: Clinical and Experimental Research*, Williams & Wilkins, 1987, **11**, 288.

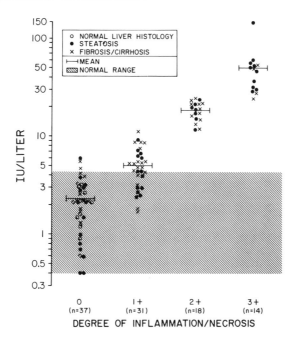

Fig. 9. Correlation of plasma GDH with the degree of liver cell necrosis in alcoholic patients undergoing liver biopsy (plasma samples obtained after 1–2 d after cessation of alcohol),[145] from "Blood Markers of Alcoholic Liver Disease" (Charles Lieber), in *Recent Developments in Alcoholism*, Vol. 6, edited by Marc Galanter (Plenum, 1988, p. 360).

Markers of Injury and Necrosis

Liver cell necrosis is considered to be an important factor in the progression of liver disease.[148] However, serum liver enzymes routinely performed as part of a screening panel, e.g., AST, ALT, and GGT, are not reliable indicators of liver necrosis.[150,154] Nouchi and co-workers (1987) reported that serum P-III-P (measured by RIA using whole antibodies) and serum laminin correlated with the degree of liver inflammation.[153] Another marker of liver injury and necrosis is plasma glutamate dehydrogenase (GDH), a liver mitochondrial enzyme predominantly located in the centrilobular area.[155] Two studies have reported that plasma GDH reflected liver necrosis as shown in Fig. 9.[150,154] Because the level of plasma GDH rapidly decays during abstinence, it should be measured within 48 h of drinking for optimal clinical utility.[154]

Conclusions

Recent advances in understanding the biochemical and physiological consequences of chronic alcohol consumption may lead to more accurate markers of alcohol consumption. Current research should lead to a battery of markers with a variety of clinical applications. The development of practical and reliable markers would lead to significant improvements in the diagnosis and treatment of alcoholism.

References

[1]B. Bush, S. Shaw, P. Cleary, T. L. Delbanco, and M. D. Aronson (1987) Screening for alcohol abuse using the CAGE questionnaire. *Am. J. Med.* **82,** 231–235.

[2]J. Persson and P. H. Magnusson (1988) Comparison between different methods of detecting patients with excessive consumption of alcohol. *Acta Med. Scand.* **223,** 101–109.

[3]P. Luchi, G. Cortis, and A. Bucarelli (1978) Forensic considerations on the comparison of "serum gammaglutamyltranspeptidase" ("gamma-GT") activity in experimental acute alcoholic intoxication and in alcoholic car drivers who caused road accidents. *Forensic Sci. Intl.* **11,** 33–39.

[4]H. Gjerde and J. Morland (1987) Concentrations of carbohydrate-deficient transferrin in dialysed plasma from drunken drivers. *Alcohol Alcoholism* **22,** 271–276.

[5]H. Orrego, L. M. Blendis, J. E. Blake, B. M. Kapur, and Y. Israel (1979) Reliability of assessment of alcohol intake based on personal interviews in a liver clinic. *Lancet* **2,** 1354–1356.

[6]J. E. Peachey and B. M. Kapur (1986) Monitoring drinking behavior with alcohol dipstick during treatment. *Alcoholism: Clin. Exp. Res.* **10,** 663–666.

[7]R. K. Fuller, K. K. Lee, and E. Gordis (1988) Validity of self-report in alcoholism research: Results of a Veterans Administration Cooperative Study. *Alcoholism: Clin. Exp. Res.* **12,** 201–205.

[8]M. Rowland and T. N. Tozer (1989) *Clinical Pharmacokinetics: Concepts and Applications* (Lea and Febiger, Philadelphia, PA), p. 385.

[9]B. J. McNeil, E. Keeler, and S. J. Adelstein (1975) Primer on certain elements of medical decision making. *N. England J. Med.* **293,** 211–215.

[10]E. L. Gottfried and E. A. Wagar (1983) Laboratory testing: A practical guide. *Disease-A-Month* **29,** 1–41.

[11]H. C. Sox (1986) Probability theory in the use of diagnostic tests. An introduction to critical study of the literature. *Annals Intern. Med.* **104,** 60–66.

[12]C. R. Cloninger (1980) Neurogenetic adaptive mechanisms in alcoholism. *Science* **236,** 410–416.

[13]C. G. Gottfries, L. Oreland, A. Wiberg, and B. Winblad (1975) Lowered monoamine oxidase activity in brains from alcoholic suicides. *J. Neurochem.* **25**, 667–673.

[14]L. Oreland, A. Wiberg, B. Winblad, C. J. Fowler, C. G. Gottfries, and K. Kiianmaa (1983) The activity of monoamine oxidase -A and -B in brains from chronic alcoholics. *J. Neural Trans.* **56**, 73–83.

[15]A. L. von Knorring, M. Bohman, L. von Knorring, and L. Oreland (1985) Platelet MAO activity as a biological marker in subgroups of alcoholism. *Acta Psych. Scand.* **72**, 51–58.

[16]G. N. Pandey, J. Fawcett, R. Gibbons, D. C. Clark, and J. M. Davis (1988) Platelet monoamine oxidase in alcoholism. *Biol. Psychol.* **24**, 15–24.

[17]J. L. Sullivan, C. N. Stanfield, S. Schanberg, and J. Cavenar (1978) Platelet monoamine oxidase and serum dopamine-β-hydroxylase activity in chronic alcoholics. *Arch. Gen. Psychol.* **35**, 1209–1212.

[18]B. A. Faraj, J. D. Lenton, M. Kutner, V. M. Camp, T. W. Stammers, S. R. Lee, P. A. Lolies, and D. Chandora (1987) Prevalence of low monoamine oxidase function in alcoholism. *Alcoholism: Clin. Exp. Res.* **11**, 464–467.

[19]B. Tabakoff, P. L. Hoffman, J. M. Lee, T. Saito, B. Willard, and F. De Leon-Jone (1988) Differences in platelet enzyme activity between alcoholics and nonalcoholics. *N. England J. Med.* **318**, 134–139.

[20]T. Ritchie, H. S. Kim, R. Cole, J. deVellis, and E. P. Noble (1988) Alcohol-induced alterations in phosphoinositide hydrolysis in astrocytes. *Alcohol* **5**, 183–187.

[21]G. Marone, M. Plaut, and L. M. Lichtenstein (1978) Characterization of a specific adenosine receptor on human lymphocytes. *J. Immunol.* **11**, 2153–2159.

[22]L. E. Nagy, I. Diamond, and A. Gordon (1988) Cultured lymphocytes from alcoholic subjects have altered cAMP signal transduction. *Proc. Natl. Acad. Sci. USA* **85**, 6973–6976.

[23]I. Diamond, B. Wrubel, W. Estrin, and A. Gordon (1987) Basal and adenosine receptor-stimulated levels of cAMP are reduced in lymphocytes from alcoholic patients. *Proc. Natl. Acad. Sci. USA* **84**, 1413–1416.

[24]E. J. Devor, T. Reich, and C. R. Cloninger (1988) Genetics of alcoholism and related end-organ damage. *Semin. Liver Disease* **8**, 1–11.

[25]K. Blum, E. P. Noble, P. J. Sheridan, A. Montgomery, T. Ritchie, P. Jagadeeswaran, H. Nogami, A. H. Briggs, and J. B. Cohn (1990) Allelic association of human dopamine D_2 receptor gene in alcoholism. *JAMA* **263**, 2055–2060.

[26]A. M. Bolos, M. Dean, S. Lucas-Derse, M. Ramsburg, G. L. Brown, and D. Goldman (1990) Population and pedigree studies reveal a lack of association between the dopamine D_2 receptor gene and alcoholism. *JAMA* **264**, 3156–3160.

[27]H. Ishii, S. Yasuraoka, Y. Shigeta, S. Takagi, T. Kamiya, F. Okuno, K. Niyamoto, and M. Tsuchiya (1978) Hepatic and intestinal gamma-glutamyltranspeptidase activity: Its activation by chronic ethanol administration. *Life Sci.* **23**, 1393–1398.

[28]G. Gadeholt, J. Aarbakke, E. Dybing, M. Sjoblom, and J. Morland (1980) Hepatic microsomal drug metabolism, glutamyl transferase activity and in vivo antipyrine half-life in rats chronically fed an ethanol diet, a control diet and a chow diet. *J. Pharmacol. Exp. Ther.* **213,** 196–203.

[29]R. Teschke, A. Brand, and G. Strohmeyer (1977) Induction of hepatic microsomal gamma-glutamyltransferase activity following chronic alcohol consumption. *Biochem. Biophys. Res. Comm.* **75,** 718–724.

[30]S. Shaw and C. S. Lieber (1980) Mechanism of increased gamma glutamyl transpeptidase after chronic alcohol consumption: Hepatic microsomal induction rather than dietary imbalance. *Substance Alcohol Actions/Misuse* **1,** 423–428.

[31]S. Yamada, J. S. Wilson, and C. S. Lieber (1985) The effects of ethanol and diet on hepatic and serum gammaglutamyltranspeptidase activities in rats. *J. Nutrition* **115,** 1285–1290.

[32]A. J. Kryszewski, G. Neale, J. B. Whitfield and D. W. Moss (1973) Enzyme changes in experimental biliary obstruction. *Clin. Chim. Acta* **47,** 175–182.

[33]A. Wu, G. Slavin, and A. J. Levi (1976) Elevated serum gammaglutamyl-transferase (transpeptidase) and histological liver damage in alcoholism. *Am. J. Gastroenterol.* **65,** 318–323.

[34]A. S. Rosman and C. S. Lieber (1991) Biological markers of alcoholism, in *Medical and Nutritional Complications of Alcoholism.* C. S. Lieber, ed. (Plenum, New York), in press.

[35]M. Salaspuro (1989) Characteristics of laboratory markers in alcohol-related organ damage. *Scand. J. Gastroenterol.* **24,** 769–780.

[36]S. Shaw, T. M. Worner, M. F. Borysow, R. E. Schmitz, and C. S. Lieber (1979) Detection of alcoholism relapse: Comparative diagnostic value of MCV, GGTP, and AANB. *Alcoholism: Clin. Exp. Res.* **3,** 297–301.

[37]M. Irwin, S. Baird, T. L. Smith, and M. Schuckit (1988) Use of laboratory tests to monitor heavy drinking by alcoholic men discharged from a treatment program. *Am. J. Psychol.* **145,** 595–599.

[38]U. J. Behrens, T. M. Worner, and C. S. Lieber (1988) Changes in carbohydrate-deficient transferrin levels after alcohol withdrawal. *Alcoholism: Clin. Exp. Res.* **12,** 539–544.

[39]A. Wu, I. Chanarin, G. Slavin, and A. J. Levi (1975) Folate deficiency in the alcoholic-its relationship to clinical and haematological abnormalities, liver disease and folate stores. *Br. J. Haematol.* **29,** 469–478.

[40]M. W. P. Carney and B. Sheffield (1978) Serum folate and B12 and haematological status of in-patient alcoholics. *Br. J. Addict.* **73,** 3–7.

[41]C. C. Okany, A. N. Bond, and S. N. Wickramasinghe (1983) Effects of ethanol on cell volume and protein synthesis in a human lymphoblastoid cell line (Raji). *Acta Haematol.* **70,** 24–34.

[42]R. J. L. Davidson and P. J. Hamilton (1978) High mean red cell volume: Its incidence and significance in routine haematology. *J. Clin. Pathol.* **31,** 493–498.

[43]H. J. Zimmerman and M. West (1963) Serum enzyme levels in the diagnosis of hepatic disease. *Am. J.Gastroenterol. 40*, 387–402.

[44]R. J. Clermont and T. C. Chalmers (1967) The transaminase tests in liver disease. *Medicine* **46**, 197–207.

[45]P. Belfrage, B. Berg, I. Hagerstrand, P. Nilsson-Ehle, H. Tornqvist, and T. Wiebe (1977) Alterations of lipid metabolism in healthy volunteers during long-term ethanol intake. *Eur. J. Clin. Invest.* **7**, 127–131.

[46]W. L. Haskell, C. Camargo, P. T. Williams, K. M. Vranizan, R. M. Krauss, F. T. Lindgren, and P. D. Wood (1984) The effect of cessation and resumption of moderate alcohol intake on serum high-density-lipoprotein subfractions. A controlled study. *N. England J. Med.* **310**, 805–810.

[47]P. Cushman, J. J. Barboriak, A. Liao, and N. E. Hoffman (1982) Association between plasma high density lipoprotein cholesterol and antipyrine metabolism in alcoholics. *Life Sci.* **30**, 1721–1724.

[48]P. V. Luoma, E. A. Sotaniemi, R. O. Pelkonen, and C. Ehnholm (1982) High-density lipoproteins and hepatic microsomal enzyme induction in alcohol consumers. *Res. Comm. Chem. Pathol. Pharmacol.* **37**, 91–96.

[49]J. Schneider, A. Liesenfeld, R. Mordasini, R. Schubotz, P. Zofel, F. Kubel, C. Vandre-Plozzitzka, and H. Kaffarnik (1985) Lipoprotein fractions, lipoprotein lipase and hepatic triglyceride lipase during short-term and long-term uptake of ethanol in healthy subjects. *Atherosclerosis* **57**, 281–291.

[50]R. Ekman, G. Fex, B. G. Johansson, P. Nilsson-Ehle, and J. Wadstein (1981) Changes in plasma high density lipoproteins and lipolytic enzymes after long-term, heavy ethanol consumption. *Scand. J. Clin. Lab. Invest.* **41**, 709–715.

[51]M. R. Taskinen, M. Valimaki, E. A. Nikkila, T. Kuusi, C. Ehnholm, and R. Ylikahri (1982) High density lipoprotein subfractions and postheparin plasma lipases in alcoholic men before and after ethanol withdrawal. *Metabolism* **31**, 1168–1174.

[52]T. Sane, E. A. Nikkila, M. R. Taskinen, M. Valimaki, and R. Ylikahri (1984) Accelerated turnover of very low density lipoprotein triglycerides in chronic alcohol users. A possible mechanism for the up-regulation of high density lipoprotein by ethanol. *Atherosclerosis* **53**, 185–193.

[53]P. Devenyi, G.M. Robinson, B. M. Kapur, and D. A. K. Roncari (1981) High-density lipoprotein cholesterol in male alcoholics with and without severe liver disease. *Am. J. Med.* **71**, 589–594.

[54]G. Heiss, N. J. Johnson, S. Reiland, C. E. Davis, and H. A. Tyroler (1980) The epidemiology of plasma high-density lipoprotein cholesterol levels. The Lipid Research Clinics Program Prevalence Study Summary. *Circulation* **62(Suppl. 4)**, 116–136.

[55]C. J. Glueck (1985) Nonpharmacologic and pharmacologic alteration of high-density lipoprotein cholesterol: Therapeutic approaches to prevention of atherosclerosis. *Am. Heart J.* **110**, 1107–1115.

[56]C. J. Glueck, P. M. Laskarzewski, D. C. Rao, and J. A. Morrison (1985) Familial aggregation of coronary risk factors, *Coronary Artery Disease: Prevention, Complication, and Treatment.* W. E. Connor and J. D. Bristow, eds. (J. B. Lippincott, Philadelphia, PA), pp. 173–192.

[57]S. Shaw and C. S. Lieber (1983) Plasma amino acids in the alcoholic: Nutritional aspects. *Alcoholism: Clin. Exp. Res.* **7,** 22–27.

[58]S. Shaw and C. S. Lieber (1980) Increased hepatic production of alpha-amino-n-butyric acid after chronic alcohol consumption in rats and baboons. *Gastroenterology* **78,** 108–113.

[59]S. Shaw and C. S. Lieber (1978) Plasma amino acid abnormalities in the alcoholic. Respective role of alcohol, nutrition, and liver injury. *Gastroenterology* **74,** 677–682.

[60]R. S. Ryback, M. J. Eckardt, B. Felsher, and R. R. Rawlings (1982) Biochemical and hematologic correlates of alcoholism and liver disease. *JAMA* **248,** 2261–2265.

[61]K. E. Freedland, M. T. Frankel, and R. C. Evenson (1985) Biochemical diagnosis of alcoholism in men psychiatric patients. *J. Stud. Alcohol* **46,** 103–106.

[62]G. Spik, B. Bayard, B. Fournet, G. Strecker, S. Bouquelet, and J. Montreuil (1975) Studies on glycoconjugates: Complete structure of two carbohydrate units of human serotransferrin. *FEBS Lett.* **50,** 296–299.

[63]H. Stibler and S. Borg (1986) Carbohydrate composition of serum transferrin in alcoholic patients. *Alcoholism: Clin. Exp. Res.* **10,** 61–64.

[64]H. Stibler and S. Borg (1981) Evidence of a reduced sialic acid content in serum transferrin in male alcoholics. *Alcoholism: Clin. Exp. Res.* **5,** 545–549.

[65]S. Petren, O. Vesterberg, and H. Jornvall (1987) Differences among five main forms of serum transferrin. *Alcoholism: Clin. Exp. Res.* **11,** 453–456.

[66]H. Stibler, O. Sydow, and S. Borg (1980) Quantitative estimation of abnormal microheterogeneity of serum transferrin in alcoholics. *Pharmacol. Biochem. Behav.* **13(Suppl. 1),** 47–51.

[67]O. Vesterberg, S. Petren and D. Schmidt (1984) Increased concentrations of a transferrin variant after alcohol abuse. *Clin. Chim. Acta* **141,** 33–39.

[68]E. L. Storey, U. Mack, L. W. Powell, and J. W. Halliday (1985) Use of chromatofocusing to detect a transferrin variant in serum of alcoholic subjects. *Clin. Chem.* **31,** 1543–1545.

[69]F. Schellenberg and J. Weill (1987) Serum desialotransferrin in the detection of alcohol abuse: Definition of a Tf index. *Drug Alcohol Depend.* **19,** 181–191.

[70]H. Stibler and S. Borg (1991) Glycoprotein glycosyltransferase activities in serum in alcohol-abusing patients and healthy controls. *Scand. J. Clin. Lab. Invest.* **51,** 43–51.

[71]H. Stibler, S. Borg, and C. Allgulander (1979) Clinical significance of abnormal heterogeneity of transferrin in relation to alcohol consumption. *Acta Med. Scand.* **206,** 275–281.

[72]A. Kapur, G. Wild, A. Milford-Ward, and D. R. Triger (1989) Carbohydrate deficient transferrin: A marker for alcohol abuse. *Br. Med. J.* **299**, 427–431.

[73]H. Stibler, S. Borg, and M. Joustra (1986) Micro anion exchange chromatography of carbohydrate-deficient transferrin in serum in relation to alcohol consumption (Swedish Patent 8400587-5) *Alcoholism: Clin. Exp. Res.* **10**, 535–544.

[74]E. L. Storey, G. J. Anderson, U. Mack, L. W. Powell, and J. W. Halliday (1987) Desialylated transferrin as a serological marker of chronic excessive alcohol ingestion. *Lancet* **1**, 1292–1294.

[75]U. J. Behrens, T. M. Worner, L. F. Braly, F. Schaffner, and C. S. Lieber (1988) Carbohydrate-deficient transferrin, a marker for chronic alcohol consumption in different ethnic populations. *Alcoholism: Clin. Exp. Res.* **12**, 427–432.

[76]I. Kwoh-Gain, L. M. Fletcher, J. Price, L. W. Powell, and J. W. Halliday (1990) Desialylated transferrin and mitochondrial aspartate aminotransferase compared as laboratory markers of excessive alcohol consumption. *Clin. Chem.* **36**, 841–845.

[77]Y. Xin, L. M. Lasker and C. S. Lieber (1991) Isoelectric focusing/ Western blotting: A novel and practical method for quantitation of carbohydrate-deficient transferrin in alcoholics. *Alcoholism: Clin. Exp. Res.* **15**, 814–821.

[78]D. I. Stott (1989) Immunoblotting and dot blotting: *J. Immunol. Methods* **119**, 153–187.

[79]A. S. Rosman, Y. Xin, K. Galvin, J. M. Lasker, E. Baraona, and C. S. Lieber, (1991) Comparison of carbohydrate deficient transferrin and red blood cell cysteine as markers of alcohol consumption. *Alcoholism: Clin. Exp. Res.* **15**, 379 (abstract).

[80]R. E. Poupon, F. Schellenberg, B. Nalpas, and J. Weill (1989) Assessment of the transferrin index in screening heavy drinkers from a general practice. *Alcoholism: Clin. Exp. Res.* **13**, 549–553.

[81]H. Stibler and R. Hultcrantz (1987) Carbohydrate-deficient transferrin in serum in patients with liver diseases. *Alcoholism: Clin. Exp. Res.* **11**, 468–473.

[82]F. Schellenberg, J. Y. Benard, A. M. Le Goff, C. Bourdin, and J. Weill (1989) Evaluation of carbohydrate-deficient transferrin compared with Tf index and other markers of alcohol abuse. *Alcoholism: Clin. Exp. Res.* **13**, 605–610.

[83]Y. Xin, J. M. Lasker, A. S. Rosman, and C. S. Lieber. (1991) Carbohydrate-deficient transferrin as measured by Western blotting: A practical yet precise marker of alcohol consumption. *Gastroenterolgy* **100**, A812 (abstract).

[84]M. A. Korsten, S. Matsuzaki, L. Feinman, and C. S. Lieber (1975) High blood acetaldehyde levels after ethanol administration. Difference between alcoholic and nonalcoholic subjects. *N. England J. Med.* **292**, 386–389.

[85]C. Di Padova, T. M. Worner, and C. S. Lieber (1987) Effect of abstinence on the blood acetaldehyde response to a test dose of alcohol in alcoholics. *Alcoholism: Clin. Exp. Res.* **11**, 559–561.

[86]C. M. Peterson and C. M. Polizzi (1987) Improved method for acetaldehyde in plasma and hemoglobin-associated acetaldehyde: Results in teetotalers and alcoholics reporting for treatment. *Alcohol* **4**, 477–480.

[87]E. Baraona, C. Di Padova, J. Tabasco, and C. S. Lieber (1987) Red blood cells: A new major modality for acetaldehyde transport from liver to other tissues. *Life Sci.* **40**, 253–258.

[88]C. M. Peterson, L. Jovanovic-Peterson, and F. Schmid-Formby (1988) Rapid association of acetaldehyde with hemoglobin in human volunteers after low dose ethanol. *Alcohol* **5**, 371–374.

[89]R. Hernandez-Munoz, E. Baraona, I. Blacksberg, and C. S. Lieber (1989) Characterization of the increased binding of acetaldehyde to red blood cells in alcoholics. *Alcoholism: Clin. Exp. Res.* **13**, 654–659.

[90]R. Uppal, A. S. Rosman, R. Hernandez, E. Baraona and C. S. Lieber (1990) Effect of liver disease on red blood cell acetaldehyde in alcoholics and non-alcoholics. *Alcoholism: Clin. Exp. Res.* **14**, 347 (abstract).

[91]C. Loguercio and G. Nardi (1987) Abnormal concentration of red blood cell cysteine in cirrhotics and its correction by S-adenosylmethionine *J. Hepatology* **5(Suppl)**, S161 (abstract).

[92]C. Alling, L. Gustavsson, and E. Anggard (1983) An abnormal phospholipid in rat organs after ethanol treatment. *FEBS Lett.* **152**, 24–28.

[93]C. Alling, L. Gustavsson, J. E. Mansson, G. Benthin, and E. Anggard (1984) Phosphatidylethanol formation in rat organs after ethanol treatment. *Biochim. Biophy. Acta* **793**, 119–122.

[94]M. F. Sorrell and D. J. Tuma (1987) The functional implications of acetaldehyde binding to cell constituents. *Ann. N.Y. Acad. Sci.* **492**, 50–62.

[95]M. A. Collins (1988) Acetaldehyde and its condensation products as markers in alcoholism. *Recent Devel. Alcohol.* **6**, 387–403.

[96]U. J. Behrens, M. Hoerner, J. M. Lasker, and C. S. Lieber (1988) Formation of acetaldehyde adducts with ethanol-inducible P450IIE1 in vivo. *Biochem. Biophys. Res. Comm.* **154**, 584–590.

[97]U. J. Behrens, X. L. Ma, S. Bychenok, E. Baraona, and C. S. Lieber (1990) Acetaldehyde-collagen adduct in CC14-induced liver injury in rats. *Biochem. Biophys. Res. Comm.* **173**, 111–119.

[98]M. A. Collins, W. P. Num, G. F. Borge, G. Teas and C. Goldfarb (1979) Dopamine-related tetrahydroisoquinolines: Significant urinary excretion by alcoholics after alcohol consumption. *Science* **206**, 1184–1186.

[99]B. Sjoquist, S. Borg, and H. Kvande (1981) Catecholamine derived compounds in urine and cerebrospinal fluid from alcoholics during and after long-standing intoxication. *Subst. Alcohol Actions / Misuse* **2**, 63–72.

[100]B. Sjoquist, S. Borg, and H. Kvande (1981) Salsolinol and methylated salsolinol in urine and cerebrospinal fluid from healthy volunteers. *Subst. Alcohol Actions / Misuse* **2**, 73–77.

[101]B. A. Faraj, V. M. Camp, D. C. Davis, D. J. Lenton, and M. Kutner (1989) Elevation of plasma salsolinol sulfate in chronic alcoholics as compared to nonalcoholics. *Alcoholism: Clin. Exp. Res.* **13**, 155–163.

[102]O. Beck, T. R. Bosin, A. Lundman, and S. Borg (1982) Identification and measurement of 6–hydroxy-1–methyl-1,2,3,4-tetrahydro–β–caboline by gas chromatography–mass spectrometry. *Biochem. Pharmacol.* **31,** 2517–2521.

[103]H. Rommelspacher, H. Damm, L. Schmidt, and G. Schmidt (1985) Increased excretion of harman by alcoholics depends on events of their life history and the state of the liver. *Psychopharmacology* **87,** 64–68.

[104]J. Adachi, Y. Mizoi, T. Fukunaga, Y. Ueno, H. Imamichi, I. Ninomiya, and T. Naito (1986) Individual difference in urinary excretion of salsolinol in alcoholic patients. *Alcohol* **3,** 371–375.

[105]F. Nomura and C. S.Lieber (1981) Binding of acetaldehyde to rat liver microsomes: Enhancement after chronic alcohol consumption. *Biochem. Biophys. Res. Comm.* **100,** 131–137.

[106]V. J. Stevens, W. J. Fanti, C. B. Newman, R. V. Sims, A. Cerami, and C. M. Peterson (1981) Acetaldehyde adducts with hemoglobin. *J. Clin. Invest.* **67,** 361–369.

[107]T. M. Donohue, D. J. Tuma, and M. F. Sorrell (1983) Acetaldehyde adducts with proteins: Binding of [^{14}C] acetaldehyde to serum albumin. *Arch. Biochem. Biophys.* **220,** 239–246.

[108]L. Lumeng and P. J. Durant (1985) Regulation of the formation of stable adducts between acetaldehyde and blood proteins. *Alcohol* **2,** 397–400.

[109]M. J. Savolainen, E. Baraona, and C. S. Lieber (1987) Acetaldehyde binding increases the catabolism of rat serum low-density lipoproteins. *Life Sci.* **40,** 841–846.

[110]Y. A. Kesaniemi, K. Kervinen, and T. A. Miettinen (1987) Acetaldehyde modification of low density lipoprotein accelerates its catabolism in man. *Eur. J. Clin. Invest.* **17,** 29–36.

[111]R. C. Lin, R. S. Smith, and L. Lumeng (1988) Detection of a protein acetaldehyde adduct in the liver of rats fed alcohol chronically. *J. Clin. Invest.* **81,** 615–619.

[112]R. C. Lin, L. Lumeng, S. Shahidi, T. Kelly, and D. C. Pound (1990) Protein-acetaldehyde adducts in serum of alcoholic patients. *Alcoholism: Clin. Exp. Res.* **14,** 438–443.

[113]X. Ma, E. Baraona, R. Hernandez-Munoz, and C. S. Lieber (1989) High levels of acetaldehyde in nonalcoholic liver injury after threonine or ethanol administration. *Hepatology* **10,** 933–940.

[114]L. B. Nguyen and C. M. Peterson (1984) The effect of acetaldehyde concentrations on the relative rates of formation of acetaldehyde-modified hemoglobin. *Proc. Soc. Exp. Bio. Med.* **177,** 226–233.

[115]K. K. Tsuboi, D. J. Thompson, E. M. Rush, and H. C. Schwartz (1981) Acetaldehyde-dependent changes in hemoglobin and oxygen affinity of human erythrocytes. *Hemoglobin* **5,** 241–250.

[116]H. D. Hoberman (1983) Post-translational modification of hemoglobin in alcoholism. *Biochem. Biophys. Res. Comm.* **113,** 1004–1009.

[117]F. R. Homaidan, L. J. Kricka, P. M. S. Clark, S. R. Jones, and T. P. Whitehead (1984) Acetaldehyde-hemoglobin adducts: An unreliable marker of alcohol abuse. *Clin. Chem.* **30**, 480–482.

[118]H. D. Hoberman and S. M. Chiodo (1982) Elevation of the hemoglobin Al fraction in alcoholism. *Alcoholism: Clin. Exp. Res.* **6**, 260–266.

[119]T. H. J. Huisman, J. B. Henson, and J. B. Wilson (1983) A new high performance liquid chromatographic procedure to quantitate hemoglobin A_{1c} and other minor hemoglobins in blood of normal, diabetic, and alcoholic individuals. *J. Lab. Clin. Med.* **102**, 163–173.

[120]T. L. Stockham and R. V. Blanke (1988) Investigation of an acetaldehyde-hemoglobin adduct in alcoholics. *Alcoholism: Clin. Exp. Res.* **12**, 748–754.

[121]R. Mobley and T. H. J. Huisman (1982) Minor hemoglobins (Hb A_1) in chronic alcoholic patients. *Hemoglobin* **6**, 79–81.

[122]C. M. Peterson, C. M. Polizzi, and P. J. Frawley (1986) Artefactual increase in hemoglobins A_{1a+b} in blood from alcoholic subjects. *Alcoholism: Clin. Exper. Res.* **10**, 219–220.

[123]E. Gordis and S. Herschkopf (1986) Application of isoelectric focusing in immobilized pH gradients to the study of acetaldehyde-modified hemoglobin. *Alcoholism: Clin. Exp. Res.* **10**, 311–319.

[124]O. Niemela, Y. Israel, Y. Mizoi, T. Fukunaga, and C. J. P. Eriksson (1990) Hemoglobin-acetaldehyde adducts in human volunteers following acute ethanol ingestion. *Alcoholism: Clin. Exp. Res.* **14**, 838–841.

[125]Y. Israel, E. Hurwitz, O. Niemela, and R. Arnon (1986) Monoclonal and polyclonal antibodies against acetaldehyde containing epitopes in acetaldehyde-protein adducts. *Proc. Natl. Acad. Sci. USA* **83**, 7923–7927.

[126]M. Hoerner, U. J. Behrens, T. M. Worner, and C. S. Lieber (1986) Humoral immune response to acetaldehyde adducts in alcoholic patients. *Res. Comm. Chem. Pathol. Pharmacol.* **54**, 3–12.

[127]O. Niemela, F. Klajner, H. Orrego, E. Vidins, L. Blendis, and Y. Israel (1987) Antibodies against acetaldehyde-modified protein epitopes in human alcoholics. *Hepatology* **7**, 1210–1214.

[128]M. Hoerner, U. J. Behrens, T. M. Worner, I. Blacksberg, L. F. Braly, F. Schaffner, and C. S. Lieber (1988) The role of alcoholism and liver disease in the appearance of serum antibodies against acetaldehyde adducts. *Hepatology* **8**, 569–574.

[129]Y. Morino, H. Kagamiyama, and H. Wada, (1964) Immunochemical distinction between glutamic-oxaloacetic transaminases from the soluble and mitochondrial fractions of mammalian tissues. *J. Biolog. Chem.* **239**, 943–944.

[130]H. Ishii, F. Okuno, Y. Shigeta, and M. Tsuchiya (1979) Enhanced serum glutamic oxaloacetic transaminase activity of mitochondrial origin in chronic alcoholics, *Currents in Alcoholism.*, vol. 5, M. Galanter, ed. (Grune and Stratton, New York), pp. 101–108.

[131]B. Nalpas, A. Vassault, S. Charpin, B. Lacour, and P. Berthelot (1986) Serum mitochondrial aspartate aminotransferase as a marker of chronic alcoholism: Diagnostic value and interpretation in a liver unit. *Hepatology* **6,** 608–614.

[132]A. Isaksson, C. Blanche, B. Hultberg, and B. Joelsson (1985) Influence of ethanol on the human serum level of betahexosaminidase. *Enzyme* **33,** 162–166.

[133]B. Hultberg, A. Isaksson, and G. Tiderstrom, (1980) Beta-hexosaminidase, leucine aminopeptidase, cystidyl aminopeptidase, hepatic enzymes and bilirubin in serum of chronic alcoholics with acute ethanol intoxication. *Clini. Chim. Acta* **105,** 317–323.

[134]P. Karkkainen, K. Poikolainen, and M. Salaspuro (1990) Serum Beta-hexosaminidase as a marker of heavy drinking. *Alcoholism: Clin. Exp. Res.* **14,** 187–190.

[135]B. Hultberg, J. H. Braconier, A. Isaksson, and L. Jansson (1981) Beta-hexosaminidase level in serum from patients with viral hepatitis as a measure of reticuloendothelial function. *Scand. J. Infect. Dis.* **13,** 241–245.

[136]B. Hultberg, A. Isaksson, and L. Jansson (1981) β-hexosaminidase in serum from patients with cirrhosis and cholestasis. *Enzyme* **26,** 296–300.

[137]B. Hultberg, A. Isaksson, B. Joelsson, A. Alwmark, P. Gullstrand, and S. Bengmark (1983) Pattern of serum beta-hexosaminidase in liver cirrhosis. *Scand. J. Gastroenterol.* **18,** 877–880.

[138]B. Hultberg and A. Isaksson (1981) A possible explanation for the occurrence of increased Beta-hexosaminidase activity in pregnancy serum. *Clin. Chim. Acta* **113,** 135–140.

[139]C. J. Waechter, C.J., and Lennarz, W.J. (1976) The role of polyprenollinked sugars in glycoprotein synthesis. *Ann. Rev. Biochem.* **45,** 95–110.

[140]R. P. Roine, U. Turpeinen, R. Ylikahri, and M. Salaspuro (1987) Urinary dolichol: A new marker of alcoholism. *Alcoholism: Clin. Exp. Res.* **11,** 525–527.

[141]R. K. Pullarkat and S. Raguthu (1985) Elevated urinary dolichol levels in chronic alcoholics. *Alcoholism: Clin. Exp. Res.* **9,** 28–30.

[142]R. P. Roine (1988) Effects of moderate drinking and alcohol abstinence on urinary dolichol levels. *Alcohol* **5,** 229–231.

[143]R. Roine, K. Humaloja, J. Hamalainen, I. Nykanen, R. Ylikahri, and M. Salaspuro (1989) Significant increases in urinary dolichol levels in bacterial infections, malignancies and pregnancy but not in other clinical conditions. *Ann. of Med.* **1,** 13–16.

[144]W. J. Powell and G. Klatskin (1968) Duration of survival in patients with Laennec's cirrhosis. Influence of alcohol withdrawal and possible effects of recent changes in general management of the disease. *Am. J. Med.* **44,** 406–420.

[145]C. S. Lieber (1988) Blood markers of alcoholic liver disease. *Recent Dev. Alcohol.* **6,** 351–365.

[146]L. Van Waes and C. S. Lieber (1977) Early perivenular sclerosis in alcoholic fatty liver: An index of progressive liver injury. *Gastroenterology* **73,** 646–650.

[147]T. M. Worner and C. S. Lieber (1985) Perivenular fibrosis as precursor lesion of cirrhosis. *JAMA* **254,** 627–630.

[148]C. S. Lieber and K. S. Guadagnini (1990) The spectrum of alcoholic liver disease. *Hosp. Pract.* **25(2A),** 51–69.

[149]S. Lieber,.L. M. DeCarli, K. N. Mak, C. Kim, and M. A. Leo (1990) Attenuation of alcohol-induced hepatic fibrosis by polyunsaturated lecithin. *Hepatology* **12,** 1390–1398.

[150]L. Van Waes and C. S. Lieber (1977) Glutamate dehydrogenase: A reliable marker of liver cell necrosis in the alcoholic. *Br. Med. J.* **2,** 1508–1510.

[151]S. Sato, T. Nouchi, T. M. Worner, and C. S. Lieber (1986) Liver fibrosis in alcoholics. Detection by Fab radioimmunoassay of serum procollagen III peptides. *JAMA* **256,** 1471–1473.

[152]E. R. Savolainen, B. Goldberg, M. A. Leo, M. Velez, and C. S. Lieber (1984) Diagnostic value of serum procollagen peptide measurements in alcoholic liver disease. *Alcoholism: Clin. Exp. Res.* **8,** 384–389.

[153]T. Nouchi, T. M. Worner, S. Sato, and C. S. Lieber (1987) Serum procollagen type III N-terminal peptides and laminin Pl peptide in alcoholic liver disease. *Alcoholism: Clin. Exp. Res.* **11,** 287–291.

[154]T. M. Worner and C. S. Lieber (1980) Plasma glutamate dehydrogenase: Clinical application in patients with alcoholic liver disease. *Alcoholism: Clin. Exp. Res.* **4,** 431–434.

[155]W. G. Guder, A. Habicht, J. Kleissl, U. Schmidt, and O. H. Wieland (1975) The diagnostic significance of liver cell inhomogeneity: Serum enzymes in patients with liver cell necrosis and the distribution of glutamate dehydrogenase in normal human liver. *Zeitschrift fur Klinische Chemie und Klinische Biochemie* **13,** 311–318.

[156]J. Chick, N. Kreitman, and M. Plant (1981) Mean cell volume and gamma-glutamyl-transpeptidase as markers of drinking in working men. *Lancet* **1,** 1249–1251.

[157]D. M. Chalmers, M. G. Rinsler, S. MacDermott, C. C. Spicer, and A. J. Levi (1981) Biochemical and haematological indicators of excessive alcohol consumption. *Gut* **22,** 992–996.

[158]G. Bliding, A. Bliding, G. Fex, and C. Tornqvist (1982) The appropriateness of laboratory tests in tracing young heavy drinkers. *Drug Alcohol Depend.* **10,** 153–158.

[159]A. G. Shaper, S. J. Pocock, D. Ashby, M. Walker, and T. P. Whitehead (1985) Biochemical and haematological response to alcohol intake. *Ann. Clin. Biochem.* **22,** 50–61.

[160]A. W. K. Chan, J. W. Welte, and R. B. Whitney (1987) Identification of alcoholism in young adults by blood chemistries. *Alcohol* **4,** 175–179.

Blood and Liver Markers in the Estimation of Alcohol Consumption

Paul Cushman, Jr.

Introduction

Despite their high rates of prevalence, problems related to alcohol consumption remain underdiagnosed.[1] The failure to recognize these problems appears to be a mixture of low index of suspicion, uncertainty, inadequate training regarding diagnostic criteria, discouragement over treatment prospects should an alcohol-related diagnosis be rendered, and patient concealment of drinking patterns. The routine inclusion of accurate biochemical tests of recent alcohol consumption in medical exams should, however, result in increased precision in recognition of such problems. Major desirable characteristics of blood markers are sensitivity, specificity, and reliability. Other important features of markers include reproducibility, stability, ease of handling, and suitability for automation.

In recent years, investigators have sought a simple and reliable test for detection of heavy alcohol use. The only infallible measure of recent alcohol use, of course, is ethanol in biological fluids. However, the presence of ethanol itself as a marker is limited by its rapid elimination rate as well as the uncertainty of how well the current level of ethanol reflects the patients' usual drinking practices. Among the long list of laboratory candidates for more useful markers, a small number

From: *Measuring Alcohol Consumption*
Eds.: R. Litten and J. Allen ©1992 The Humana Press Inc.

have merited sustained clinical and epidemiological interest. This review will focus primarily on liver enzymes and red blood cell size as markers of alcohol use.

γ-Glutamyl Transpeptidase (Transferase)

Gamma-glutamyl transpeptidase (GGTP) is largely a hepatic, membrane-bound peptide. Its serum levels are easily measured by automated multichannel analyzers. GGTP concentrations are elevated by ethanol, liver disease, several enzyme inducers, and possibly tobacco smoking, as well as by other drugs and diseases.[2,3] The mechanisms by which ethanol increases GGTP levels are still unclear, but the most likely explanation is enzyme induction. Reduced biliary excretion of GGTP, augmented release of GGTP from the liver because of early ethanol toxicity, and reduced hepatic uptake of GGTP have also been implicated.[3,4] Apparently, serum levels are mainly independent of GGTP enzyme activity in the liver itself.[5]

Marker retests of reliability and stability of GGTP are at acceptable levels. Serum GGTP normalizes in hospitalized alcoholics during supervised abstinence in 2–6 wk and exhibits a half-life of about 2 wk. Since hospitalized patients with nonethanol-associated liver disease usually do not show significant changes in serum GGTP over the short term, documentation of decreases in serum GGTP in hospitalized liver disease patients can prove quite helpful in distinguishing alcoholics from others with nonalcohol-related liver disease.[3] The sensitivity of GGTP in hospitalized alcoholics is acceptably high, especially in those with manifest liver disease. Specificity, however, is more problematic. In hospitalized alcoholics with little or only minor liver disease, GGTP sensitivity tends to be lower but specificity higher. GGTP sensitivities in ambulatory treatment-seeking alcoholics tend to be in the 40–50% range.[3,6–10] These data support a major if not critical role for hepatic dysfunction as a causal mechanism between ethanol and elevated serum GGTP.

Several investigators have evaluated the practical utility of serum GGTP as a potential marker for alcohol disorder levels in healthy, ambulatory individuals.[11–15] In addition, several research projects have compared self-report histories of recent alcohol use with quantitative serum GGTP levels. The preponderance of the data suggests that GGTP is quite useful in distinguishing those reporting heavy drinking (i.e., 50 g/d) from others. In a series of studies involving 11,643 males in Malmo, Sweden, Kristenson et al. (1980) and Kristenson and Trell (1982) reported a GGTP sensitivity of 0.33 as a measure of heavy drinking habits.[12,14] Six percent of the middle-aged men subset had high levels of GGTP; three quarters of them also reported heavy drinking. Whitfield et al. reported GGTP sensitivity of 25% and specificity of 94% in a sample of approximately 4000 males surveyed to detect heavy daily drinking.[6]

Clark et al. (1983) found similar sensitivity and specificity (50 and 92%, respectively) for GGTP in healthy government employees consuming in excess of 70 drinks/a week.[15] Contrary to these positive relationships, Isacsson et al. (1987) reported rather questionable utility of GGTP for detection of alcohol-abuse cases in another large Swedish cohort.[16] Likewise, Penn and Worthington (1983) were similarly skeptical about GGTP as a screen for heavy alcohol.[17] Only 50% of their patients with high GGTP were subsequently determined to be heavy drinkers.

Despite difficulties in self-quantification of alcohol consumption, several studies have offered a possible relationship between GGTP and personal estimation of alcohol consumption. Devgun et al. (1985) studied serum GGTP after heavy but unstandardized acute doses of ethanol in a large group of normals.[18] Despite high blood–alcohol levels, no corresponding change in serum GGTP ensued. On the other hand, Freer and Statland (1977) were able to document a noticeable rise in GGTP after repeated dosing of twice weekly pulses of 0.75 mg/kg ethanol.[19] Eighty-two of the normal males had significantly increased GGTP levels by the third or fourth week. In an independent project in which daily quantitative ethanol levels in urine were submitted by alcoholics in ambulatory treatment of severe liver disease, Orrego et al. (1985) found a strong correlation between ethanol intake and GGTP levels.[20] The apparently abstinent patients were defined by the absence of ethanol in almost daily morning urine submitted for weeks. Their serum GGTP levels diverged sharply downward from those whose urine revealed evidence of continued drinking. Kristenson et al. (1983) performed a long-term followup on a group of 500 male alcoholics identified by high serum GGTP levels.[21] Those randomly assigned to alcohol treatment exhibited lower GGTP

Table 1
γ-GGTP as a Marker for Alcohol Disorders

Population studies	Sensitivity, %	Specificity, %
Hospital patients, general acute care		
Manifest liver disease	75–90	8–25
Little or mild liver disease	36–50	85–90
Hospital patients, alcohol dependency/abuse recognized		
Manifest liver disease	80–92	
Little or mild liver disease	46–60	
Ambulatory community population*	13–50	75–94

*Only 3–16% positive for serum GGTP above the normal range.

levels, lower quantities of consumption, and lesser alcohol-related problems than did the untreated sample.

Several investigators have assessed GGTP as a quantitative index of alcohol consumption across a spectrum of unselected persons. Despite the failure of Robinson et al. (1979) to relate GGTP levels to quantified recent alcohol use in 260 males, several other researchers have found a positive, albeit rather low, correlation between consumption and GGTP levels.[11,13–15,22–27] Results of these studies are summarized in Table 1. Interestingly, gender seems to modify the relationship. For example, Papoz et al. (1981) found correlation coefficients of 0.35 for men and 0.24 for women in 995 healthy outpatients.[28] At high levels of alcohol consumption (above 60 g/d), only a rough correlation with GGTP was found in Swedish outpatient dependent pregnant females.[29] Similar conclusions were reached by Bell and Steensland (1987) in studies of Norwegians of both genders.[30]

As previously noted, GGTP can provide useful corroborative data in studies where drinking level is a dependent variable.[20,21] Further support of GGTP as a blood marker for alcohol has been offered by Wallace et al. (1988) based on their followup study of patients with modest alcohol problems.[31] Given brief advice vs no special intervention, the brief advice group reported lower quantities of ethanol consumption in both 6- and 12-mo followups and revealed small but significant reductions in serum GGTP. Also, Puddey et al. (1988) discovered that normal nonproblem drinking subjects lowered their levels of GGTP during periods of reduced alcohol consumption.[32] Declines in ethanol consumption from 335 to about 70 mL/wk were associated with mean GGTP decrease from 34 to 29 g/L. In the same experiment, modest decreases in mean corpuscular volume (MCV) and serum lipoproteins were also associated with lowered alcohol consumption.

Mean Corpuscular Volume

MCV is widely available as a part of the routine Coulter Counter blood-testing procedure. It is elevated in alcoholism as well as in folic acid and vitamin B_{12} deficiencies, reticulocytoses, and after use of certain drugs. The mechanism of ethanol-related macrocytosis is not understood. Plasma, liver, and red cell folate levels are often normal, though sometimes low, and do not correlate with MCV values.[33,34] Furthermore, Wu et al. (1974) have demonstrated that folate treatment does not correct macrocytosis in chronic alcoholics.[33] An ethanol effect on some red cell stem cell precursor has been postulated.

As a marker for alcohol disorders, MCV has considerable utility. Its sensitivity among hospital patients is usually quite high, and it appears to be little influenced by presence or absence of manifest liver disease.[3,7,9,10,15,35] Its specificity for alcohol disorders is much higher than GGTP. Ambulatory alcohol-treatment centers report sensitivity in the 0.25–0.50 range. MCV's sensitivity, as with GGTP, appears to be lower in females than males.[7,9,28,36,37] However, in surveys of general, unselected patients in ambulatory clinics, MCV seems inefficient as a marker for alcohol disorders. For example, Unger and Johnson (1974) surveyed 8000 cases and found only 3%—most of whom were heavy alcohol users—had a high level of MCV.[38] Whitfield et al. reported similar results in 7915 Australians, as did Whitehead et al. in the United Kingdom.[6,11]

As reflected in Table 2, reported rates of sensitivity and specificity for MCV tend to be in the same range as those found with GGTP for comparable samples. Asker et al. (1982) believed a 1.7fl increase in MCV corresponded to each 10 g of ethanol consumed daily.[39] Other researchers have found substantially higher variabilities across individuals, and it appears that the relationship between MCV and level of consumption does not assume a simple linear form. As with GGTP, MCV can be very effective in separating extreme heavy drinkers from those drinking little. Distinguishing moderately high-consuming individuals, however, is difficult.

Table 2
MCV as a Marker for Alcohol Disorders

Population studies	Sensitivity, %	Specificity, %
Hospital patients, general acute care		
Manifest liver disease	45–88	80–90
Little or mild liver disease	42–72	65–99
Hospital patients, alcohol dependency/abuse recognized		
Both mild or manifest liver disease	45–88	
Ambulatory, alcohol dependency/abuse		
Both mild and manifest liver disease	21–45	
Ambulatory community sample*	21–75	40–64

*Only 3–6% positive for MCV values above the normal range.

Other Hepatic Enzymes and Lipids

Other candidate markers for problematic drinking include the aminotransferases. Serum glutamic oxaloacetic transaminase (SGOT) or aspartate aminotransferase (AST) is elevated in a variety of nonalcohol-related liver diseases, cardiac and skeletal muscle disorders, alcohol-related liver disease. With abstinence, alcoholics with initially high serum AST levels show erratic fluctuations of AST. This aminotransferase is generally viewed as too insensitive and nonspecific for clinical utility in the identification of alcohol disorders.[15,25,36,40] Apparently, considerable liver damage must occur before AST exceeds normal ranges. Quantity of drinking generally correlates quite poorly with the level of consumption measured by questionnaire.[11,25,40] Recently, the mitochondrial AST/total AST ratio in serum has been proposed as a more effective marker for alcoholism.[41] Further research on this ratio would be most welcome.

Serum alkaline phosphatase (AP) may also increase in a variety of medical conditions (e.g., bone disorders, growth, drugs, and nonalcohol-related liver disturbances) as well as in alcohol abuse. However, AP is widely regarded as too nonspecific and insensitive to aid in the detection of alcohol disorders.[15,25,27] Glutamate dehydrogenase (GDH) levels may bear some relationship with alcohol consumption but appear to be primarily reflective of liver disease.[28,42] However, low Spearman coefficients—much lower than with MCV or GGTP—and the rapid change in GDH levels after the cessation of drinking limit its utility as a marker of alcohol consumption.[26]

An interesting association with high-density lipoproteins (HDL) and ethanol has often been observed.[31,43–45] In particular, the data seem to support a meaningful relationship between plasma high-density lipoprotein-cholesterol (HDL-C) concentration and the quantity of ethanol consumed. A relative increase in serum HDL-C of about 1 mg/d may correspond to about an ounce of ethanol consumed weekly.[45] Unfortunately, clinical experience indicates that fasting serum levels of HDL-C fall in a broad range, are age- and sex-related, and are affected by tobacco, exercise, estrogens, and severe liver disease, as well as by certain drugs. Sizable alcohol loads may be necessary to move HDL-C above normal limits. The half-life of HDL-C elevation seems to be approximately 7 d in abstinent alcoholics.[44] Current research suggests that it is the lipoprotein rather than the cholesterol component of HDL (A-I and A-II) that elevates with ethanol feeding.[32,46,47] The relationship of lipoprotein levels to quantities of recent alcohol consumption clearly warrants further exploration. In general, the sensitivity of the lipoproteins seems too weak to warrant their use in detection of alcoholism and consumption patterns.[47] Changes in lipoproteins against a known baseline might, however, prove a useful index of change in drinking behaviors in clinical studies of alcoholism treatment effectiveness.

Combinations of Traditional Markers

Since individual markers have proven somewhat disappointing, attention has recently been directed toward using markers in combination. Theoretically, markers that are independently influenced by different tissues to ethanol would be especially desirable, e.g., combining red cell MCV with liver GGTP and HDL-C might be more revelatory than either alone.[9,27,45]

In hospital-based populations, alcoholics with liver diseases usually reveal very high levels of GGTP. Including other measures tends to raise their combined sensitivity but lower their specificity. Nevertheless, many researchers have found that pooling markers, such as

GGTP, MCV, and AP, is useful in accurately distinguishing alcoholics from other gastroenterological and nonalcoholic liver disease patients.[36] Sanchez-Craig and Annis (1981) found that HDL-C and GGTP in combination were more effective in discriminating abstinent or light drinkers from moderate to heavy drinkers.[43] Combined markers correctly classified 62.5% of the patients, whereas GGTP and HDL-C singly identified only 50 and 42%, respectively. On the contrary, Pol et al. (1990) reported that combining GGTP and MCV was not more effective than using GGTP alone in discriminating hospitalized alcoholics, with or without alcohol liver disease, from nonalcoholic liver disease patients.[3]

Recently, considerable interest has focused on elaborate computer statistical modeling of data obtained from multiple routine liver and hematological values.[48] Ryback et al. (1982) and Ryback et al. (1982) applied quadratic discriminant analysis to 25 standard blood tests.[49,50] The optimal predictor combination of scores correctly classified 100% of nonalcoholics without liver disease, 95% of alcoholics with evident liver disease, 98% of alcoholics without evident liver disease, and 89% of nonalcoholics with liver disease. This finding is promising, but cross-validation is needed.[48] Beresford and colleagues (1990) contrasted the CAGE with computer-assisted laboratory data manipulations in 915 unselected hospitalized patients.[51] Using separate interviews leading to the DSM-III-R diagnosis, the authors distinguished 244 alcohol-dependent and 671 nondependent persons. The CAGE questionnaire separately was 76% sensitive and 94% specific. Linear and quadratic discriminant functions of standard blood chemistries (GGTP excluded), however, were unable to enhance its validity.

Particularly among ambulatory alcoholic samples, pooled markers appear more sensitive than single markers in isolation.[3,9,27,28] For example, in a large trial of outpatient dependent alcoholics vs social drinkers, Skinner et al. (1984) found that a combination of GGTP, HDL-C, MCV, and current blood-alcohol level had a 0.71:0.90 sensitivity–specificity ratio.[52] The comparable ratio for family practice patients was 0.84:0.50. Biochemical data were less accurate than clinical signs or medical history, which included an alcohol-suggestive trauma component. Similarly, sensitivity of single and pooled markers fell considerably short of the CAGE questionnaire among general hospital patients.[53] Similar results were reported by Beresford et al. (1990).[51]

Table 3
Correlation Coefficients Between Reported Alcohol Consumption
and Biological Markers in Men

Reference	N	GGTP	MCV
23	238	0.35	—
24	2152	0.36	0.27
13	488	0.31–0.42	0.36–0.44
28	604	0.35	0.34
6	1503	0.36	0.39
43	40	0.46	—
27	54	0.09–0.16	0.20–0.27
54*	126	0.20	0.33

*Lee and DeFrank (1988) only includes 71% females, 29% males.

Using pooled variables in estimations of self-reported alcohol consumption among nonalcohol populations has also been disappointing. Lee and DeFrank (1988) reported mixed results on GGTP, MCV, MAST, and CAGE with university staff and students.[54] The number of drinks each week correlated with MCV, CAGE, and GGTP at 0.33, 0.23, and 0.20, respectively (Table 3). On the other hand, Papoz and colleagues (1981) found a rough correlation between drinking estimates and GGTP and MCV, both singly and pooled using multiple regressions.[28] The more sensitive GGTP could be profitably supplemented by MCV. Chick et al. (1981) found that both markers produced few false-positives and had low sensitivity for detection of alcohol use of 0.45 g/wk in their study of 488 workers in a British alcohol-producing business.[13]

Conclusions

Several hepatic and blood measures have been considered as potential markers for alcohol disorders and recent alcohol consumption. The better ones, MCV and GGTP, have shown greatest sensitivity for alcohol disorders among hospitalized alcoholics and less sensitivity with ambulatory, treatment-seeking alcoholics. Only modest utility has been reported for unselected, healthy individuals. Sensitivities tend to be lower for females than males for both GGTP and MCV. Pooled markers generally are superior to single ones, although interview instruments frequently continue to fare even better. For estimation of recent consumption in ambulatory alcohol or unselected patients, liver and blood markers are of some help, especially in distinguishing heavy vs very low categories of subjects.

References

[1]R. D. Moore, L. R. Bone, G. Geller, J. A. Mamon, E. J. Stokes, and D. M. Levine (1989) Prevalence, detection and treatment of alcoholism in hospitalized patients. *JAMA* **261**, 403–407.

[2]E. Reyes and W. R. Miller (1980) Serum gamma-glutamyl transpeptidase as a diagnostic aid in problem drinkers. *Addict. Behav.* **5**, 59–65.

[3]S. Pol, T. Poynard, P. Bedossa, S. Naveau, A. Aubert, and J.-C. Chaput (1990) Diagnostic value of serum γ-glutamyltransferase activity and mean corpuscular volume in alcohol patients with or without cirrhosis. *Alcoholism: Clin. Exp. Res.* **14**, 250–254.

[4]R. Barouki, M. N. Chobert, J. Finidori, M. Aggerbeck, B. Nalpas, and J. Hanoune (1983) Ethanol effects in rat hepatoma cell line: Induction of gamma glutamyl transferase. *Hepatology* **3**, 323–329.

[5]M. J. Sellinger, D. S. Matloff, and M. M. Kaplan (1982) γ-Glutamyl transpeptidase activity in liver disease: Serum elevation is independent of hepatic GGTP activity. *Clin. Chim. Acta* **125**, 283–290.

[6]J. B. Whitfield, W. J. Hensley, D. Bryden, and G. Gallagher (1978) Some laboratory correlates of drinking habits. *Ann. Clin. Biochem.* **15**, 297–303.

[7]R. M. Morse and R. D. Hurt (1979) Screening for alcoholism. *JAMA* **242**, 2688–2690.

[8]R. B. Garvin, D. W. Foy, and G. S. Alford (1981) A critical examination of GGTP as a biochemical marker for alcohol abuse. *Addict. Behav.* **6**, 377–383.

[9]P. Cushman, G. Jacobson, J. J. Barboriak, and A. Anderson (1984) Biochemical markers for alcoholism: Sensitivity problems. *Alcoholism: Clin. Exp. Res.* **8**, 253–257.

[10]R. R. Watson, M. E. Mohs, C. Eskelson, R. E. Sampliner, and B. Hartmann (1986) Identification of alcohol abuse and alcoholism with biological parameters. *Alcoholism: Clin. Exp. Res.* **10**, 364–385.

[11]T. P. Whitehead, A. Clark, and A. G. W. Whitfield. (1978) Biochemical and haematological markers of alcohol intake. *Lancet* **1**, 978–981.

[12]H. Kristenson, E. Trell, G. Fex, and B. Hood (1980) Serum gamma-glutamyl transferase: Statistical distribution in a middle aged population and evolution of drinking habits in individuals with elevated levels. *Prevent. Med.* **9**, 108–119.

[13]J. Chick, N. Kreitman, and M. Plant (1981) Mean cell volume and gamma-glutamyl-transpeptidase as markers of drinking in working men. *Lancet* **1**, 1249–1251.

[14]H. Kristenson and E. Trell (1982) Indicators of alcohol consumption: Comparisons between a questionnaire (MmMAST), interviews, and serum gamma-glutamyl transferase (GGT) in a health survey of middle aged men. *Brit. J. Addict.* **77,** 297–304.

[15]P. M. S. Clark, R. Holder, M. Mullet, and J. P. Whitehead (1983) Sensitivity and specificity of lab tests for alcohol abuse. *Alcohol Alcoholism* **18,** 261–269.

[16]S.-O. Isacsson, B. S. Hanson, L. Janzon, S.-E. Lindell, and B. Steen (1987) Methods to assess alcohol consumption in 68 year old men: Results from a population study in Sweden. *Brit. J. Addict.* **82,** 1235–1244.

[17]R. Penn and D. J. Worthington (1983) Is serum GGTP a misleading test? *Brit. Med. J.* **286,** 531–532.

[18]M. S. Devgun, J. A. Dunbar, J. Hagart, B. T. Martin, and S. A. Ogsten (1985) Effect of acute and varying amounts of alcohol consumption in alkaline phosphatase, AST and GGTP. *Alcohol Alcoholism* **9,** 235–236.

[19]D. E. Freer and B. E. Statland (1977) Effects of ethanol on the activities of selected enzymes in sera of healthy young adults. *Clin. Chem.* **23,** 1099–1102.

[20]H. Orrego, J. E. Blake, and Y. Israel (1985) Relationship between γ-GTP and mean urinary alcohol levels in alcoholics while drinking and after alcohol withdrawal. *Alcoholism: Clin. Exp. Res.* **9,** 10–13.

[21]H. Kristenson, H. Ohlin, M.-B. Hutlen-Nosslin, E. Trell, and B. Hood, B. (1983) Identification and intervention of heavy drinking in middle aged men: Results and followup of 24–60 months of long term study with randomized controls. *Alcoholism: Clin. Exp. Res.* **7,** 203–209.

[22]D. Robinson, C. Monk, and G. Bailey (1979) Relationship between serum gamma-glutamyl transpeptidase and reported alcohol consumption in healthy men. *J. Stud. Alcohol* **40,** 896–899.

[23]J. G. Rollason, G. Pincherle, and D. Robinson (1972) Serum gamma glutamyl transpeptidase in relation to alcohol consumption. *Clin. Chim. Acta* **39,** 75–80.

[24]A. Bagrel, A. d'Houtand, R. Gueguen, and G. Siest (1979) Relations between reported alcohol consumption and certain biological variables in an "unselected" population. *Clin. Chem.* **25,** 1242–1246.

[25]C. Gluud, I. Anderson, O. Dietrichson, B. Gluud, A. Jacobsen, and E. Johl (1981) GGTP, aspartate aminotransferase and alkaline phosphatase as markers of alcohol consumption in OPD alcoholics. *Eur. J. Clin. Invest.* **11,** 171–176.

[26]M. W. Bernadt, J. Numford, C. Taylor. B. Smith, and R. M. Murray (1982) Comparison of questionnaire and laboratory tests in the detection of excessive drinking and alcoholism. *Lancet* **1,** 325–328.

[27]R. Poikolainen, P. Karkkainen, and J. Pikkarainen (1985) Correlations between biological markers and alcohol intake as measured by diary and questionnaire in men. *J. Stud. Alcohol* **46,** 383–387.

[28]L. Papoz, J.-M. Narnet, G. Pequinot, E. Eschwege, J. R. Claude, and D. Schwartz (1981) Alcohol consumption in a healthy population: Relationship to γ-glutamyl transferase activity and mean corpuscular volume. *JAMA* **245,** 1748–1751.

[29]C. Hollstedt and L. Dahlgren (1987) Peripheral markers in female "hidden alcoholic." *Acta Psych. Scand.* **75**, 591–596.

[30]H. Bell and H. Steensland (1987) Serum activity of gamma-glutamyltranspeptidase (GGT) in relation to estimated alcohol consumption and questionnaires in alcohol dependence syndrome. *Brit. J. Addict.* **82**, 1021–1026.

[31]P. Wallace, S. Cutler, and A. Haines (1988) Randomized controlled trial of general practitioner intervention in patients with excessive alcohol consumption. *Brit. Med. J.* **297**, 663–668.

[32]I. B. Puddey, J. R. L. Masarei, R. Vandongen, and L. J. Berlin (1988) Serum apolipoprotein A-II as a marker of change in alcohol intake in male drinkers. *Alcohol Alcoholism* **21**, 375–383.

[33]A. Wu, I. Charanin, and A. J. Levy, (1974) Macrocytosis of chronic alcoholism. *Lancet* **1**, 829–831.

[34]D. Savage and J. Lindenbaum (1986) Anemia in alcoholics. *Medicine* **65**, 322–331.

[35]D. M. Chalmers, I. Chanarin, S. Macdermott, and A. J. Levi (1980) Sex-related differences in the hematological effects of excessive alcohol consumption. *J. Clin. Pathol.* **33**, 3–7.

[36]D. M. Chalmers, G. M. Rinsler, S. MacDermott, C. C. Spicer, and A. J. Levi (1981) Biochemical and haemotological indicators of excessive alcohol consumption. *Gut* **22**, 992–996.

[37]H. A. Skinner, S. Holt, W. J. Sheu, and Y. Israel (1986) Clinical versus laboratory detection of alcohol abuse: The alcohol clinical index. *Brit. Med. J.* **292**, 1703–1708.

[38]K. W. Unger and D. Johnson (1974) Red blood cell mean corpuscular volume: A potential indicator of excessive alcohol usage in a working population. *Am. J. Med. Sci.* **267**, 281–289.

[39]R. L. Asker, J. H. Renwick, and A. H. Goldstone (1982) Erythrocyte volume as a crude indicator of ethanol consumption in pregnancy. *Clin. Lab. Haematol.* **4**, 115–119.

[40]S. Holt, H. A. Skinner, and Y. Israel (1981) Early identification of alcohol abuse: Clinical and laboratory indicators. *Can. Med. Assn. J.* **124**, 1279–1294.

[41]B. Nalpas, A. Vassault, S. Charpin, B. Lacour, and P. Berthelot (1984) Serum activity of mitochondrial aspartate aminotransferase: A sensitive marker of alcoholism with or without alcoholic hepatitis. *Hepatology* **4**, 893–896.

[42]F. Schellenberg, J. Neill, and J. N. Lamy (1983) Place de la glutamate deshydrogenase dans la depistage biologique des buveurs excessifs. *Annales De Biologie Clinique* **41**, 256–282.

[43]M. Sanchez-Craig and H. M. Annis (1981) GGTP and high density lipoprotein cholesterol in male problem drinkers: Advantages of a composite index for predicting alcohol consumption. *Alcoholism: Clin. Exp. Res.* **5**, 540–545.

[44]P. Cushman, J. J. Barboriak, and J. Kalbfleish (1986) Alcohol: High density lipoproteins, apolipoproteins. *Alcoholism: Clin. Exp. Res.* **10**, 154–157.

[45]R. D. Moore and T. A. Pearson (1986) Moderate alcohol consumption and coronary artery disease. *Medicine* **65,** 242–266.

[46]Y. Okamoto, Y. Fujimori, H. Nakano, and T. Tsujii (1988) The role of the liver in the alcohol induced high-density lipoprotein metabolism. *J. Lab. Clin. Med.* **111,** 482–485.

[47]P. Puchois, T. Fonyan, J.-L. Gentilini, P. Gelez, and J.–C. Fruchart (1989) Serum apoliproprotein AII: A biochemical indicator of alcohol abuse. *Clin. Chim. Acta* **185,** 185–189.

[48]K. E. Freeland, M. T. Frankel, and R. C. Evenson (1985) Biochemical diagnosis of alcoholism in men psychiatric patients. *J. Stud. Alcohol* **46,** 103–106.

[49]R. S. Ryback, M. J. Eckhardt, B. Felsher, and R. R. Rawlings (1982) Biochemical and hematological correlates of alcoholism and liver disease. *JAMA* **248,** 2261–2265.

[50]R. S. Ryback, M. J. Eckhardt, R. R. Rawlings, and L. S. Rosenthal (1982) Quadratic discriminant analysis as an aid to interpretive reporting of clinical laboratory tests. *JAMA* **248,** 2342–2345.

[51]T. P. Beresford, F. C. Blow, R. Singer, and M. R. Lucey (1990) Comparison of CAGE questionnaires and computer assisted laboratory profiles in screening for covert alcoholism. *Lancet* **2,** 482–484.

[52]H. A. Skinner, S. Holt, R. Schuller, J. Roy, and Y. Israel (1984) Identification of alcohol abuse using laboratory tests and a history of trauma. *Ann. Internal Med.* **101,** 847–850.

[53]B. Bush, S. Shaw, P. Cleary, T. L. Delbanco, and M. D. Aronson (1987) Screening for alcohol abuse using the CAGE questionnaire. *Am. J. Med.* **82,** 231–237.

[54]D. J. Lee and R. S. DeFrank (1988) Interrelationships between self report alcohol consumption, physiological indices and alcoholism screening measures. *J. Stud. Alcohol* **49,** 532–536.

Carbohydrate-Deficient Transferrin and 5-Hydroxytryptophol

Two New Markers of High Alcohol Consumption

Stefan Borg, Olof Beck, Anders Helander, Annette Voltaire, and Helena Stibler

Introduction

Successful treatment of alcoholism is generally dependent on an early diagnosis. Consequently, there is a clinical need for objective and reliable markers of alcohol consumption, since valid information of alcohol intake cannot be obtained from subjects misusing alcohol. This chapter describes two new biochemical markers of alcohol consumption: carbohydrate-deficient transferrin (CDT) in serum and 5-hydroxytryptophol (5-HTOL) in urine. It gives information about biochemical background and clinical characteristics and describes recent method developments to enable their use in clinical routine work. Because they show different clinical profiles, how they can be used in combination in a clinical setting monitoring alcohol consumption is also illustrated.

From: *Measuring Alcohol Consumption*
Eds.: R. Litten and J. Allen ©1992 The Humana Press Inc.

Carbohydrate-Deficient Transferrin

Fifteen years ago, it was reported that the microheterogeneity of serum transferrin was altered in more than 80% of alcohol-abusing patients and that the transferrin pattern normalized during abstention.[1-3] Evaluation of the transferrin pattern in patients with various nonalcohol-related diseases, including liver diseases, failed to show the transferrin abnormality unless there was a history of excessive alcohol consumption.

The first studies were carried out using isoelectric focusing (IEF). When analyzed by this method, the microheterogeneity of normal human diferric serum transferrin of the predominant phenotype C is largely dependent on the number of negatively charged terminal sialic acid residues in the carbohydrate chains of transferrin. The main component with an isoelectric point (pI) of 5.4 contains 4 moles of sialic acid per mole of transferrin (tetrasialotransferrin).[4-6]

Normal serum also contains small amounts of transferrin with higher pI values considered to represent tri- and disialotransferrin (pI 5.6 and 5.7, respectively) as well as components with lower pIs probably corresponding to penta- and hexasialotransferrin (pI 5.3 and 5.2, respectively).[4-6] The abnormality in alcoholic patients consisted of elevated amounts of transferrin with pI 5.7 and to a lesser degree, with pI 5.8 and 5.9, representing di-, mono-, and asialotransferrin, respectively.[1-3,7] Studies of the carbohydrate composition of transferrin have shown that transferrin from alcoholic patients contained significantly lower concentrations of sialic acid than normal and that the contents of neutral galactose and N-acetylglucosamine were also reduced.[7,8] These findings resulted in the denomination carbohydrate-deficient transferrin, or CDT.

The IEF technique originally used was regarded as too complex for routine purposes. Therefore, a simplified method was developed based on isocratic anion-exchange chromatography in disposable microcolumns followed by a transferrin radioimmune assay (RIA).[9] The first technique to be used measured as CDT all transferrin components that were isoelectric at above pH 5.65. The assay showed a coefficient of variation between 5 and 9%, a sensitivity of 91%, and a specificity of 99% when comparing 100 alcoholic patients with 115 healthy, "normal consumers" and 40 total abstainers. Elevated levels were noted in healthy individuals after daily consumption of 60 g of ethanol during a 10-d period. In a sample of 344 patients with nonalcohol-related conditions, there was a 2% false-positive rate. Treatment with a number of commonly used drugs was not associated with an elevation of CDT values. The correlation between CDT and reported daily alcohol consumption during the preceding month in 80 healthy, "normal consumers" and 77 alcoholic patients was 0.21 and 0.22, respectively. During abstinence following alcohol consumption, the mean half-life of elevated CDT values was calculated to be 17 d.[9-12]

These findings were soon confirmed by several other independent groups employing various methods.[13-20] In one study, alcoholics were followed as outpatients for up to 6 mo during long-term, abstinence-oriented treatment. In some cases (10%), elevated levels of CDT were reported despite these patients' denial of consumption of alcohol. Therefore, the authors discussed the possibility of other reasons for the rise in CDT values in alcoholic patients.[21]

The original CDT assay had the disadvantage that correct separation depended on a buffer of very low ionic strength with a stable pH of 5.65. To increase the technical stability, a modification was introduced: The anion-exchange chromatography of serum was changed so that it was based on ionic strength rather than on pH. Thereby, mainly isotransferrins with a pI greater than 5.7 were measured.[22] Clinical evaluation of 251 individuals showed results similar to the original CDT assay. However, since the more cathodal isotransferrins were principally measured, the reference levels were lower (Table 1). The sensitivity for alcohol abuse was 93% and specificity 98% in the groups studied. The correlation between reported, total current alcohol consumption and serum CDT was 0.38 ($p < 0.01$). Slightly elevated false-positive levels were seen in single cases of primary biliary cirrhosis, chronic active hepatitis, and drug-induced liver injury. False-positive results have also been found in individuals with rare, genetic D-variants of transferrin.[23] In patients with a newly discovered inborn error of glycoprotein metabolism (CDG syndrome), extremely high CDT levels are constantly present.[24,25]

Table 1
Results from Different Subject Groups with Modified CDT Assay[22]

Group	CDT (mg/L)	
	Mean ± SD	Range
Healthy "normal consumers", $n = 71$		
Males ($n = 43$)	13 ± 2	8–17
Females ($n = 28$)	17 ± 3	11–22
Alcohol-abusing patients, $n = 78$		
Males ($n = 58$)	44 ± 25	9–136
Females ($n = 20$)	32 ± 13	13–60
Abstaining alcoholic patients, $n = 20$	15 ± 2	10–16
Patients with liver disease, $n = 55$		
Patients without current abuse,		
$n = 51$	12 ± 5	2–32
Patients with current abuse,		
$n = 4$	22 ± 5	16–29
Patients with CDG syndrome,		
$n = 27$	240 ± 81	110–416

5-Hydroxytryptophol

A natural constituent of human urine, 5-HTOL can be used as a marker of recent alcohol consumption (within 24 h). It is formed in the body as a metabolite of serotonin (5-hydroxytryptamine, or 5-HT) and is excreted in urine after conjugation with glucuronic acid.[26] The catabolic pathway of serotonin consists of oxidative deamination to 5-hydroxyindole-3-acetaldehyde (5-HIAL) by the enzyme monoamine oxidase (EC 1.4.3.4). 5-HIAL is then either oxidized by aldehyde dehydrogenase (EC 1.2.1.3) to form 5-hydroxyindole-3-acetic acid (5-HIAA) or reduced by alcohol dehydrogenase (EC 1.1.1.1) or aldehyde reductase (EC 1.1.1.2) to form 5-HTOL.

Normally, 5-HIAA is by far the most abundant end product of 5-HT metabolism.[27] However, the catabolism is altered shortly after alcohol intake, resulting in decreased formation of 5-HIAA and a concomitant dose-dependent increase in 5-HTOL. This effect is suggested to be the result of either competitive inhibition of oldehyde dehydrogenase by the ethanol-derived acetaldehyde or the increased ratio between NADH and NAD$^+$, both of which favour reduction of 5-HIAL to 5-HTOL.[28]

The shift toward the reductive pathway is observable in the urine several hours after the ethanol has disappeared from the body.[29] Based on these findings, increased urinary excretion of 5-HTOL is presently being evaluated as a possible diagnostic marker of recent alcohol consumption in the treatment of alcohol dependence. To compensate for urine dilution, ratios of 5-HTOL to 5-HIAA and 5-HTOL to creatinine have been used.

So far, urinary 5-HTOL (free + conjugated form) has been determined by a sensitive and specific gas chromatographic-mass spectrometric (GC-MS) method.[29,30] One important improvement that has been made is the use of bacterial β-glucuronidase for the hydrolysis of conjugated 5-HTOL. With the new enzyme, the hydrolysis is performed in 1 h as opposed to 16 with the previously used enzyme from *Helix pomatia* (sulfatase type H-1). However, mass spectrometric methods are expensive and therefore not suitable when large numbers of samples are to be analyzed routinely. Recently, a simple and reproducible high-performance liquid chromatography (HPLC) method for the determination of elevated urinary levels of 5-HTOL has been developed.[31] This method is probably more suitable for use in clinical laboratory screening. In brief, a 20- µL sample was hydrolyzed for 1 h at 37°C using β-glucuronidase. A centrifuged aliquot was then diluted with ice-cold deionized water and injected directly into an isocratically eluted Nucleosil C_{18} reversed-phase column.

Alternatively, 5-HTOL was isolated by a one-step sample cleanup procedure on a small Sephadex G10 column prior to HPLC analysis. After the sample cleanup, no major interfering peaks were observed in the chromatograms. Presence of 5-HTOL was determined by electrochemical detection (+ 0.60 V vs Ag/AgCl reference). A good correlation ($r^2 = 0.97$) was obtained between the HPLC method and the GC-MS method when analyses were performed on urine samples containing 200–2500 pmol/mL of 5-HTOL. The intraassay coefficients of variation with a 5-HTOL standard containing 1000 pmol/mL and a urine sample containing 1800 pmol/mL were 1.4 and 1.7%, respectively ($n = 6$). In urine (as a conjugate), 5-HTOL was stable for at least 12 mo when stored at –20 or –80°C.

For routine clinical use, urinary 5-HTOL is preferably expressed relative to 5-HIAA instead of creatinine, since ingestion of foods rich in 5-HT (e.g., bananas, pineapple, kiwi, tomatoes, and walnuts) could

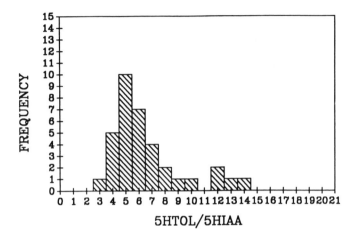

Fig. 1. Distribution of 5-HTOL/5-HIAA values in urine from 30 inpatients in a medical ward during long-term abstention from alcohol. Data is expressed as a molar ratio of 5-HTOL to 5-HIAA × 1000.

otherwise yield false-positive results.[32] This also contributes to increased sensitivity of the marker, since 5-HTOL increases at the expense of 5-HIAA during ethanol metabolism. Urinary 5-HIAA is conveniently determined by an HPLC method based on direct sample injection as described in detail elsewhere.[33]

A limit to discriminate between a normal and an elevated value of the 5-HTOL to 5-HIAA molar ratio has been set after studying groups of people during periods of abstinence from alcohol (Fig. 1). A value of 0.02 of the 5-HTOL to 5-HIAA molar ratio was selected as a suitable cut-off. At this limit, the probability of having a positive outcome in the test during abstinence from alcohol was calculated to be less than 0.001.[30] In a group of four volunteers drinking over 200 g of ethanol a week, more than 50% of the urine samples collected during a 3-mo period were greater than 0.02 in the 5-HTOL/5-HIAA assay. To date, the only known factors other than alcohol consumption that can lead to an elevated 5-HTOL to 5-HIAA ratio are disulfiram and calcium cyanamide therapy (to be published). Disulfiram appeared to be more effectual in elevating the 5-HTOL to 5-HIAA ratio than calcium cyanamide.

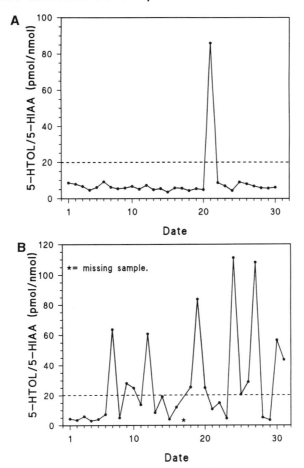

Fig. 2. The molar ratio of 5-HTOL to 5-HIAA × 1000 in urine from two alcoholic patients during a treatment period of 1 mo.

More than 20 patients have been followed during treatment with this marker. Examples from two patients are shown in Fig. 2. In the first patient (Fig. 2A), one occasion with an abnormal 5-HTOL to 5-HIAA ratio was detected, and the consumption of alcohol was verified clinically. In this patient, the stability of the 5-HTOL to 5-HIAA ratio during periods of abstinence from alcohol is apparent. In the other example (Fig. 2B), the patient had relapsed into regular alcohol consumption, which was revealed by elevated 5-HTOL to 5-HIAA ratios.

Table 2
Frequency of Relapses in Alcohol Consumption in 15 Patients*

Frequency of positive indications of total observations	Number of patients with positive indication			
	Self-report	CDT	5-HTOL	CDT and/or 5-HTOL
Frequent' > 10%	3	5	4	6
Some, 1–10%	5	3	11	9
None, 0%	7	7	—	—

*Patients' goal was to stay abstinent for 6 mo. Lapses were verified by self-report during personal investigation three times a week, CDT once a week, or daily urinary 5-HTOL sent to the laboratory by the patient.

Clinical Use of Both Markers

The urinary ratio of 5-HTOL to 5-HIAA was used to validate CDT values and clinical information on alcohol intake from 15 alcoholic patients attempting to remain abstinent for 3–6 months (to be published). CDT in serum was determined once a week, whereas urine was collected daily for 5-HTOL and 5-HIAA analysis. The results are summarized in Table 2. During the treatment period, elevated levels of CDT were noted in six cases. In no case did CDT rise without confirmation of alcohol intake by clinical information and/or by elevated 5-HTOL/5-HIAA levels. Interestingly, 5-HTOL/5-HIAA levels revealed that a number of patients were drinking alcohol in a controlled way from time to time without admitting to alcohol consumption. Six patients had elevated 5-HTOL/5-HIAA as the only indicator of alcohol consumption. No subject appeared to remain totally abstinent from alcohol during the observation period.

Conclusions

In summary, the modified version of the CDT assay seems to have the same clinical properties as the original method. By validating the drinking history with urinary 5-HTOL/5-HIAA, CDT was shown to be highly consistent over time, and in all cases with an elevated value of CDT, alcohol consumption could be verified. Furthermore, urinary 5-HTOL shows properties of becoming a valuable marker of recent alcohol consumption.

References

[1]H. Stibler and K. G. Kjellin (1976) Isoelectric focusing and electrophoresis of the CSF proteins in tremor of different origins. *J. Neurol. Sci.* **30**, 269–285.

[2]H. Stibler, S. Borg, and C. Allgulander (1979) Clinical significance of abnormal heterogeneity of transferrin in relation to alcohol consumption. *Acta Med. Scand.* **206**, 275–281.

[3]H. Stibler, O. Sydow, and S. Borg (1980) Quantitative estimation of abnormal microheterogeneity of serum transferrin in alcoholics. *Pharmacol. Biochem. Behav.* **13**, 47–51.

[4]A. Hamann (1977) Microheterogeneity of serum glycoproteins as revealed by flat-bed isoelectric focusing, *Electrofocusing and Isotachophoresis.* B. J. Radfola and B. Graesslin, eds. (W de Gruyter, New York), pp. 329–335.

[5]H. Stibler (1978) The normal cerebrospinal fluid proteins identified by means of thin-layer isoelectric focusing and crossed immunoelectro-focusing. *J. Neurol. Sci.* **36**, 273–288.

[6]H. van Eijk, W. van Noort, M.-L. Dubelaar, and C. van der Heul (1983) The microheterogeneity of human transferrin in biological fluids. *Clin. Chim. Acta.* **132**, 167–171.

[7]H. Stibler and S. Borg (1986) Carbohydrate composition of transferrin in alcoholic patients. *Alcohol. Clin. Exp. Res.* **10**, 61–64.

[8]H. Stibler and S. Borg (1981) Evidence of a reduced sialic acid content in serum transferrin in male alcoholic patients. *Alcohol. Clin. Exp. Res.* **5**, 545–549.

[9]H. Stibler, S. Borg, and M. Joustra (1986) Micro anion exchange chromatography of carbohydrate-deficient transferrin in serum in relation to alcohol consumption. *Alcohol. Clin. Exp. Res.* **10**, 535–544.

[10]H. Stibler and R. Hultcrantz (1987) Carbohydrate-deficient transferrin in serum in patients with liver diseases. *Alcohol. Clin. Exp. Res.* **11**, 468–473.

[11]H. Stibler and S. Borg (1988) The value of carbohydrate-deficient transferrin as a marker of high alcohol consumption, *Biomedical and Social Aspects of Alcohol and Alcoholism.* K. Kuriyama, A. Takada, and H. Ishii, eds. (Elsevier Sci., Amsterdam), pp. 503–506.

[12]H. Stibler, L. Dahlgren, and S. Borg (1988) Carbohydrate-deficient transferrin (CDT) in serum in women with early alcohol addiction. *Alcohol* **5**, 393–398.

[13]O. Vesterberg, S. Petren, and D. Schmidt (1984) Increased concentrations of a transferrin variant after alcohol abuse. *Clin. Chim. Acta.* **141**, 33–39.

[14]E. Storey, U. Mack, L. Powell, and J. Halliday (1985) Use of chromatofocusing to detect a transferrin variant in serum of alcoholic subjects. *Clin. Chem.* **31**, 1543–1545.

[15]F. Schellenberg and J. Weill (1987) Serum desialotranserrin in the detection of alcohol abuse. *Drug Alcohol Depend.* **19**, 181–191.

[16]H. Gjerde, J. Johnsen, A. Bjorneboe, G.-E. Bjorneboe, and J. Morland (1988) A comparison of serum carbohydrate-deficient transferrin with other biological markers of excessive drinking. *Scand. J. Clin. Lab. Invest.* **48**, 1–6.

[17]U. Behrens, T. Worner, L. Brady, F. Schaffner, and C. Lieber (1988) Carbohydrate-deficient transferrin: A marker for chronic alcohol consumption in different ethnic populations. *Alcohol. Clin. Exp. Res.* **12**, 427–432.

[18]F. Schellenberg, J. Benard, A. Le Goff, C. Bourdin, and J. Weill (1989) Evaluation of carbohydrate-deficient transferrin compared with Tf index and other markers of alcohol abuse. *Alcohol. Clin. Exp. Res.* **13**, 605–610.

[19]A. Kapur, G. Wild, A. Milford-Ward, and D. R. Triger (1989) Carbohydrate-deficient transferrin: A marker for alcohol abuse. *Brit. Med. J.* **299**, 427–431.

[20]I. Kwoh- Galn, L. Fletcher, J. Price, L. Powell, and J. Halliday (1990) Desialylated transferrin and mitochondrial aspartate aminotransferase compared to laboratory markers of excessive alcohol consumption. *Clin. Chem.* **36**, 841–845.

[21]U. Behrens, T. Worner, and C. Lieber (1988) Changes in carbohydrate-deficient transferrin levels after alcohol withdrawal. *Alcohol. Clin. Exp. Res.* **12**, 539–544.

[22]H. Stibler, S. Borg, and M. Joustra (1991) A modified method for the assay of carbohydrate-deficient transferrin (CDT) in serum. *Alcohol Alcoholism* **Suppl. 1**, 451–454.

[23]H. Stibler, S. Borg, and G. Beckman (1988) Transferrin phenotype and level of carbohydrate-deficient transferrin in serum in healthy individuals. *Alcohol. Clin. Exp. Res.* **12**, 450–453.

[24]H. Stibler and J. Jaeken (1990) Carbohydrate-deficient serum transferrin in a new hereditary systemic syndrome. *Arch. Dis. Child.* **65**, 107–111.

[25]H. Stibler, J. Jaeken, and B. Kristiansson (1991) Biochemical characteristics and diagnosis of the carbohydrate-deficient glycoprotein syndrome. *Acta Paediatr. Scand.* **Suppl. 375**, 22–31.

[26]T. R. Bosin (1978) Serotonin metabolism, *Serotonin in Health and Disease,* vol. 1, W. B. Essman, ed. (Spectrum, New York), pp. 181–300.

[27]O. Beck, S. Borg, G. Jonsson, A. Lundman, and P. Valverius (1984) Measurement of 5-hydroxytryptophol and 5-hydroxyindoleacetic acid in human and rat brain and plasma. *J. Neural Transm.* **59**, 57–67.

[28]M. J. Walsh (1973) Role of acetaldehyde in the interactions of ethanol with neuroamines. *Adv. Mental Sci.* **3**, 233–266.

[29]O. Beck, S. Borg, L. Eriksson, and A. Lundman (1982) 5-Hydroxytryptophol in the cerebrospinal fluid and urine of alcoholics and healthy subjects. *Naunyn-Schmiedebergs Arch. Pharmacol.* **321**, 293–297.

[30]A. Voltaire, O. Beck, and S. Borg (1992) Urinary 5-hydroxytryptophol: A possible marker of recent alcohol consumption. *Alcohol. Clin. Exp. Res.* **16**, (in press).

[31]A. Helander, O. Beck, and S. Borg (1991) HPLC determination of urinary 5-hydroxytryptophol: A biochemical marker of recent alcohol consumption. *Alcohol. Clin. Exp. Res.* **15**, 386 (abstract).

[32]A. Helander, T. Wikström, C. Löwenmo, G. Jacobsson, and O. Beck (1992) Urinary excretion of 5-hydroxyindole-3-acetic acid and 5-hydroxytryptophol after oral loading with serotonin. *Life Sci.* (in press).
[33]A. Helander, O. Beck, M. Wennberg, T. Wikström, and G. Jacobsson (1991) Determination of urinary 5-hydroxyindole-3-acetic acid by high-performance liquid chromatography with electrochemical detection and direct sample injection. *Anal. Biochem.* **196,** 170–173.

Protein–Acetaldehyde Adducts as Biochemical Markers of Alcohol Consumption

Lawrence Lumeng and Renee C. Lin

Introduction

There are two types of diagnostic markers for alcohol abuse and alcoholism: trait (or vulnerability) markers and state markers.[1,2] Trait markers include indicators that will identify individuals who are more vulnerable than the general population to develop alcoholism; and as the term implies, they are tests that elicit certain genetic traits. By comparison, state markers are tests that will identify alcohol abuse by reflecting psychosocial, biochemical, or physiological changes brought about by chronic and excessive alcohol intake. Because of the enormous problem of alcohol abuse and alcoholism in the West, research to devise a test or a panel of tests to serve as trait or state markers is critically important.

Since this book is devoted to techniques to assess alcohol consumption, only state markers will be discussed. The needs to develop sensitive and accurate state markers of alcohol abuse are obvious. First, they are needed so that the diagnosis of alcohol abuse can be made early, when alcohol drinking first becomes a problem and before deleterious consequences become apparent. Second, they will

From: *Measuring Alcohol Consumption*
Eds.: R. Litten and J. Allen ©1992 The Humana Press Inc.

be very useful in monitoring abstinence in alcoholic patients undergoing outpatient alcoholism treatment or followup. Third, they will be very useful to physicians who are responsible to confirm their clinical impression that their patient's illness, e.g., cirrhosis, pancreatitis, or gastritis, is caused by alcohol abuse. Lastly, they will be essential for clinical research in the alcoholism field particularly for epidemiologic and treatment evaluation research.

An ideal state marker for detection of excessive alcohol intake should be one that is highly sensitive and specific. It should exhibit high predictive accuracy and not be affected by factors unrelated to alcoholism, e.g., age, sex, smoking, drug use (particularly agents such as disulfiram, benzodiazepines, or anticonvulsants), malnutrition, and concomitant diseases. The ideal marker should also be able to distinguish social from heavy drinkers and diagnose both binge and nonbinge drinkers. The ideal marker should also be noninvasive, simple, convenient, and easy to automate; thus, it can be adopted by hospital laboratories. Additionally, it should remain abnormal for at least a few days after cessation of alcohol ingestion yet sensitive enough to quickly detect relapse in drinking. Unfortunately, currently, we do not have such a test.

Current tests for detecting alcohol abuse include several structured interviews and questionnaires, a battery of older, easily available, conventional marker tests, and a number of newly proposed tests that are in the research stage. The two most popular questionnaires are the Michigan Alcoholism Screening Test (MAST) and CAGE.[3,4] These questionnaires are structured to detect psychopathological syndromes that accompany alcohol abuse and to elicit behavioral sequelae that are the earliest signs of problem drinking. They have been shown to be more sensitive and specific than currently available conventional laboratory tests in detecting alcohol abuse, however, their usefulness is limited by patient denial and physician hesitancy to use them.[5] Conventional laboratory tests are listed in Table 1, and newer markers that are still in the research stage are listed in Table 2.

Studies on the newer markers thus far have been limited to comparison between known alcoholics and nonalcoholics and the effects of the presence or absence of liver disease. It is expected that the sensitivity and specificity of each of these newer tests will decrease when it is used to screen the general population in which there are

Table 1
Older, Conventional State Markers
of Excessive Alcohol Intake*

Aspartate aminotransferase (AST)
Alanine aminotransferase (ALT)
γ-Glutamyl transferase (GGT)
α-Amino-*n*-butyric acid/leucine
Glutamate dehydrogenase (GDH)
RBC mean corpuscular volume (MCV)
HDL-cholesterol (HDL-C)
Uric acid (UA)
Ferritin

*Modified from Chan (1990).[2]

Table 2
Newer State Markers
of Excessive Alcohol Intake Still Under Investigation*

Tests	Reference
Urinary dolichol	65
Serum acetate	66
Serum 2,3-butanediol	67
Urinary alcohol-specific products	68
Blood acetaldehyde adducts	46–48
Mitochondrial aspartate aminotransferase	69
RBC aldehyde dehydrogenase	70
Apolipoprotein AII	71
Transferrin variants (CDT)	72

*Modified from Chan (1990).[2]

drinkers with different levels of alcohol intake. The newer markers
can be divided into two categories: those that are directly linked to
alcohol metabolism—e.g., urinary dolichol, serum acetate, serum 2,3-
butanediol, red blood cells (RBC) aldehyde dehydrogenase, urinary
alcohol-specific products, and blood acetaldehyde adducts (AAs)—

and those that are more indirectly related to alcohol metabolism— e.g., mitochondrial aspartate aminotransferase (mAST), apolipoprotein AII, and transferrin variants [carbohydrate-deficient transferrin (CDT)].

Because of the proximity of the former group of new markers to the pathway of alcohol metabolism, one might speculate that they will be more sensitive and specific than the more indirect tests, but this remains speculative. In fact, measurements of serum acetate and 2,3-butanediol may have very little value as screening tests because their detection is dependent on the presence of ethanol or the need to give a test dose of alcohol. In theory, measurements of urinary alcohol-specific products (presumably breakdown products of protein-AAs) and blood AAs may reflect the presence of elevated blood acetaldehyde levels from alcohol drinking integrated over time. Therefore, one might speculate that these measurements can be used in a way analogous to the use of glycated hemoglobin (Hb A1C) or albumin in diabetic patients to monitor the degree of glycemia control as a function of time. However, there are still many technical problems in assaying urinary alcohol-specific products and blood AAs. The subject of this paper will concentrate on blood AAs, particularly protein-AAs in venous blood.

Alcohol and Acetaldehyde Metabolism

After ingestion, only a small percentage of alcohol is excreted unchanged in breath and urine; most of it is metabolized by way of enzymatic oxidation.[6] The principal pathway of ethanol oxidation involves its conversion to acetaldehyde and then to acetate. The major organ responsible for ethanol oxidation is the liver. Alcohol dehydrogenase (ADH) and a microsomal ethanol oxidizing system (MEOS or P450IIE1) catalyze the conversion of ethanol to acetaldehyde. A mitochondrial isoform of aldehyde dehydrogenase (ALDH2), which has a low K_m for acetaldehyde, mediates the further oxidation of acetaldehyde to acetate.[6-8] Because the capacity of ALDH2 to metabolize acetaldehyde is quite high, acetaldehyde, once formed, is rapidly eliminated. In individuals who do not abuse alcohol, ADH is quantitatively the most important enzyme responsible for the oxidation of ethanol. However, since P450IIE1 is inducible by chronic alcohol use, this microsomal oxidation system will contribute more to ethanol metabolism in chronic alcoholics than nonalcoholics.[9]

Accurate measurement of blood and tissue acetaldehyde is still very difficult because of its volatility, rapid elimination, chemical reactivity, and artifactual generation from ethanol.[10,11] With improved methods, blood acetaldehyde levels in the range of 1–50 μM have been reported in humans consuming alcohol.[12] Other than methodological problems, the sources of variation in blood acetaldehyde concentrations include genetically determined differences in the isoforms of ADH and ALDH2, environmental factors (such as chronic alcohol abuse), liver injury, and exogenous and endogenous sources of acetaldehyde other than alcohol ingestion.[12–15] Using flush reaction measured by a laser-doppler skin probe as an indirect indicator of blood acetaldehyde levels and by measuring alcohol elimination rates, recent studies by Thomasson et al. (1990) showed clearly that Asians with the $\beta_2\beta_2$ ADH phenotype (or ADH2^2 homozygote genotype) eliminated alcohol faster and developed a more intense flush reaction than those with $\beta_1\beta_2$ ADH phenotype (or heterozygote ADH2^1–ADH2^2 genotype).[16] Additionally, among Asians with $\beta_1\beta_2$ ADH phenotype, the alcohol elimination rate was slower and flush reaction more severe in those with inactive ALDH or (ALDH2^2 genotype) than those with the active ALDH (or ALDH2^1 genotype). The slower rate of alcohol elimination in Asians with the ALDH2^2 genotype is probably explained by product inhibition of ADH because of acetaldehyde accumulation. Although blood acetaldehyde levels were not measured in these experiments, these data indicate that genetically determined differences in ADH and ALDH2 isoenzyme patterns are likely to play an important role in explaining the interindividual differences in the rates of acetaldehyde formation and elimination in vivo.

There is no question that the major source of acetaldehyde in vivo comes from ingestion of ethanol; however, some evidence now indicates that there may be other exogenous and endogenous sources. Gastrointestinal microflora can convert fermentable ethanol and carbohydrates to acetaldehyde, and absorption of acetaldehyde from the gut can be increased in individuals with bacterial overgrowth syndromes.[17,18] Additionally, several enzyme-catalyzed reactions in vivo in theory can generate endogenous acetaldehyde. These enzyme-catalyzed reactions may include pyruvate dehydrogenase, threonine aldolase, deoxypentosephosphate aldolase, and phosphorylethanolamine phospholyase.[19–22]

Chemical Properties of Acetaldehyde and Chemistry of AA Formation

Acetaldehyde is a highly volatile liquid that is readily soluble in water and organic solvents. It diffuses rapidly across cell membranes and is chemically much more reactive than ethanol. The electrophilic nature of its carbonyl carbon makes it susceptible to attack by a variety of nucleophilic groups, especially free amino groups, e.g., ε-amino group of N-lysine and N-terminus amino group of proteins.

Currently, the chemistry of AA formation is inadequately understood. Acetaldehyde can form unstable adducts with $-NH_2$, $-SH$, guanido- and imidazolo-groups of proteins. When it reacts with amino groups, the unstable modification mainly exists in the form of Schiff's bases. These Schiff's bases can follow several chemical fates: For one, Schiff's bases can dissociate back to acetaldehyde and protein; or else, they can undergo an exchange reaction with another amino group. Additionally, Schiff's bases can undergo stabilizing reactions, i.e., reduction reaction to form reduced Schiff's bases that are irreversible; nucleophilic addition of thiols or another acetaldehyde molecule, formation of imidazolidinone derivatives, and so on.[23] Finally, acetaldehyde has been shown to form cross-links with macromolecules such as proteins and DNAs.

Formation of Protein-AA (Including Hemoglobin-AA) In Vitro

Acetaldehyde, even at low concentrations, has been shown to form adducts in vitro with lipids, nucleic acids, and proteins. Among proteins, the following have been shown to react with acetaldehyde in vitro: plasma proteins,[24] albumin,[25] erythrocyte membrane proteins,[26] hepatic microsomal proteins,[27] and a number of enzymes and functional proteins including tubulin,[28] actin,[29] RNase A,[30] and most importantly, hemoglobin.[23] In many of these proteins, such as tubulin, there seem to be certain key lysine residues that show unusual reactivity toward acetaldehyde.[31] The reason for the enhanced reactivity of these key lysines remains poorly understood, but it most likely depends on the particular microenvironment in which these lysines

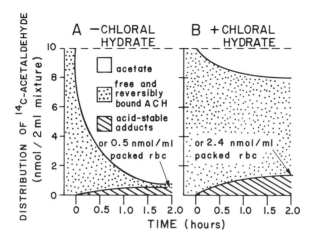

Fig. 1. The fate of ^{14}C-acetaldehyde (5 μ*M*) incubated with 2 mL of diluted human venous blood at 37°C in the absence (Panel A) and presence (Panel B) of chloral hydrate.

reside.[32] For instance, the local microenvironment can lower the pKa of the amino group, and it can bring about participation of neighboring functional groups to promote adduct formation.

In the case of hemoglobin, Stevens et al. (1981) were the first to report that acetaldehyde, at 3 μ*M*–3 m*M*, reacted with hemoglobin to form both stable and unstable adducts when added in vitro.[33] When borohydride was added to a mixture of ^{14}C-acetaldehyde and hemoglobin, they found that the ^{14}C-acetaldehyde reacted covalently with valine, lysine, and tyrosine residues of globin. Furthermore, they demonstrated by chromatography that the reaction of acetaldehyde with hemoglobin Ao (HbAo) produced modified hemoglobins that migrated in the HbA$_{1a-c}$ region.

Since the initial report by Stevens et al. (1981), a number of studies have confirmed that acetaldehyde in the 5–20 μM range can react in vitro with hemoglobin as well as plasma proteins to form AAs. In our laboratory, 5 μM ^{14}C-acetaldehyde was incubated with 2 mL of diluted human venous blood and in the absence or presence of chloral hydrate to partially inhibit acetaldehyde oxidation by RBCs (Fig. 1).[34] After a 2-h incubation, we found that in the presence of chloral hydrate, 7.8 of the 10 nmol of ^{14}C-acetaldehyde originally added to the diluted blood remained unmetabolized. Of the amount unmetabolized, 25% formed irreversible protein-AAs (not dialyzable, not extractable with semicarbazide, and acid stable), and 75% remained free (60%) or formed reversible adducts (15%).

In the absence of chloral hydrate, there was considerable metabolism of ^{14}C-acetaldehyde, i.e., only 0.8 of the 10 nmol of ^{14}C-acetaldehyde added remained unmetabolized. By Affi-gel blue affinity chromatography, it was observed that all the ^{14}C-acetaldehyde adducts in the plasma fraction eluted in three peaks, one of which was albumin. By Biorex-70 chromatography, it was found that almost all the ^{14}C-acetaldehyde adducts in RBC lysate were associated with the fast-moving fractions of hemoglobin. Borohydride reduction was not required to produce the irreversible protein-AAs found in plasma and RBC lysate.

In addition to work in our laboratory, Nguyen and Peterson (1986) separated the subunits of hemoglobin by p-hydroxymercurybenzoate treatment and cation-exchange chromatography after reacting hemoglobin with ^{14}C-acetaldehyde.[35] They found that the ratio of specific radioactivities of β- to α-chain was at least eight, indicating that acetaldehyde added in vitro preferentially reacted with the β-chain.

San George and Hoberman (1986) were able to confirm the work of Nguyen and Peterson.[23] However, in addition, these workers showed by ^{13}C-NMR analysis that acetaldehyde reacts with the amino termini of hemoglobin as well as synthetic polypeptides to form stable cyclic imidazolidinone derivatives. Our laboratory has also examined the rates of reaction of acetaldehyde (5 μM) with oxy-HbAo, deoxy-HbAo, and carbonyl-HbAo, as well as the effects of pyridoxal-5'-P and ascorbic acid.[24] We found that acetaldehyde reacted more rapidly with deoxy-HbAo than oxy- or carbonyl-HbAo and that pyridoxal-5-P inhibited whereas ascorbic acid facilitated the irreversible binding of acetaldehyde to hemoglobin.

Our laboratory and others have developed high-performance liquid chromatographic (HPLC) methods to separate human hemoglobin into more than 13 peaks.[36–38] By using a Synchropak CM300 cation-exchange column, we found that incubation of 10 μM acetaldehyde with 88 μM carbonyl-HbAo at 37°C for 5 h led to the formation of three fast-moving Hb peaks, i.e., 3', 4, and 11, representing 0.04, 0.06,

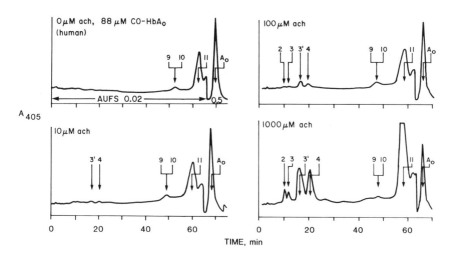

Fig. 2. Formation of three fast-moving hemoglobin peaks detected by Synchropak CM300 cation-exchange column HPLC when varying concentrations (0–1000 μM) of acetaldehyde was incubated with 88 μM of carbonyl-HbAo at 37°C for 5 h.

and 0.18%, respectively, of the total carbonyl-HbAo (Fig. 2). These three fast-moving peaks increased in percentage as a function of acetaldehyde concentrations. This work has been confirmed recently by Sillanaukee and Koivula (1990), who also used a strong cation exchanger (MonoS HR 5/5) and attributed two peaks, HbAlach and HbAlc, to adduct formation with acetaldehyde.[38] These two peaks also increased in peak height as a function of acetaldehyde concentrations (from 10 to 5000 μM). Additionally, they observed that when hemoglobin was incubated with 1000 μM of [1-2 ^{14}C]acetaldehyde, the radioactivities bound to HbAlach and HbAlc were ten- and five-fold higher, respectively, than that in HbAo. More recently, studies by Gapstur and associates (1991) have further corroborated these earlier findings.[39] The latter investigators reported that with erythrocytes incubated in the presence of cyanamide (to inhibit acetaldehyde oxidation) and 5 μM ^{14}C-acetaldehyde, about 0.1% of the hemoglobin became ^{14}C-labeled in 3 h and 9% became modified in 48 h. Peptide maps of modified hemoglobin further revealed several sites where stable modifications could be detected.

Formation of Protein-AA in Liver In Vivo

In addition to experiments in vitro with incubation of acetaldehyde with erythrocytes and lysates, a number of other studies now clearly indicate that acetaldehyde can form stable adducts with proteins in vivo.[40–42] As summarized in Table 3, these studies include:

1. Finding circulating antibodies directed against protein-AAs in mice fed alcohol chronically and in alcoholics with and without liver disease;[43–45]
2. The detection of several liver protein-AAs in rats exposed to ethanol chronically, i.e., the 37 kDa protein-AA, P450IIE1-AA; [40,41] and
3. The detection of two serum protein-AAs and hemoglobin-AAs in experimental animals exposed to alcohol and in alcoholic patients.

The latter group of studies will be discussed in the next two sections.[46–48] Of note, collagen I-AA has been found in rats treated with CCl_4, and it presumably arises from endogenous acetaldehyde.[42]

To date, the most extensively studied protein-AA found in vivo is the 37 kDa liver protein-AA detected in rats fed alcohol chronically and in primary cultured rat hepatocytes exposed to ethanol.[40,49–51] The method of detection involved SDS-PAGE and immunotransblot using antihemocyanin–AA IgG and antimyoglobin–AA IgG, which can recognize AAs as epitopes. By this method and based on a series of experiments in vivo and in vitro, several important characteristics of this 37 kDa protein-AA have been discovered:

1. The 37 kDa liver protein–AA is detected in rats as early as within 1 wk of alcohol feeding. It was found by feeding rats either the high-fat Lieber-DeCarli or a low-fat AIN'76 alcohol-containing liquid diet.[49]
2. Feeding rats an alcohol-containing liquid diet supplemented with cyanamide (a specific inhibitor of the low km ALDH) raised plasma acetaldehyde concentrations more than sixfold and greatly increased the intensity of the 37 kDa protein-AA band on immunotransblot.[49]
3. The bond between the 37 kDa protein and acetaldehyde is very stable, and its stability was not enhanced by adding cyanoborohydride.[40]
4. The 37 kDa liver protein-AA is not a newly synthesized hepatic protein.[40]
5. Based on immunotransblot, the 37 kDa liver protein–AA did not react with antibodies monospecific for rat ADH or rat cytosolic ALDH.

Table 3
Formation of Protein-AAs In Vivo During Alcohol Exposure

Circulating antibodies against protein-AAs detected in mice fed ethanol and in alcoholics
Detection of several protein-AAs in liver of rats fed alcohol chronically: 37kDa protein-AA and P450IIE1-AA
Detection of two serum protein-AAs and hemoglobin-AA in experimental animals exposed to alcohol and in alcoholics

Thus, this protein–AA is not a chemically modified ADH or ALDH or their degradation fragments.[49]

6. On subcellular fractionation, the 37 kDa protein–AA was found in cytosol. However, in experiments in which cyanamide was added to the alcohol-containing liquid diet, the 37 kDa protein–AA was found both in cytosol and the plasma membrane fraction as well.[52] Recently, this data, was further confirmed by flow cytometry.

7. In primary cultured rat hepatocytes exposed to 5 mM of ethanol, the 37 kDa liver protein–AA could be detected within 3 d of alcohol exposure.[51] The extent of formation of this protein–AA was dependent on the steady-state concentrations of ethanol and acetaldehyde.

8. Formation of the 37 kDa protein–AA in cultured hepatocytes was completely abolished in the presence of 10 μM of 4-methylpyrazole. Since this concentration of 4-methylpyrazole is one-tenth of the known K_i value of 4-methylpyrazole for cytP450IIE1, it was concluded that ADH but not cytP450IIE1 is responsible for supplying the acetaldehyde needed to form the 37 kDa protein–AA.[51]

Measurement of Hemoglobin–AA In Vivo

Table 4 summarizes the approaches that have been tried to measure the formation of hemoglobin–AAs in animal and clinical studies. These approaches have included: *(1)* ion-exchange chromatography,[33,36,37,53,54] isoelectric focusing (IEF),[55] and affinity column[56]; *(2)* fluorigenic HPLC[46]; and *(3)* immunologic approaches using either antihemocyanin–AA and antimyoglobin–AA IgGs[47] or antibovine serum albumin–AA IgG that had been cross-absorbed with human RBC proteins coupled to CNBr–Sepharose 4B and further purified by human RBC protein–AA Sepharose 4B.[48]

Table 4
Approaches to Measuring Hemoglobin–AA(HB-AA)
in Clinical and Animal Studies

Ion-exchange chromatography, IEF, and affinity column

Fluorogenic HPLC, i.e., 1,3-cyclohexanedione

Immunologic Approaches

  Anti-BSA-AA cross-absorbed with human RBC proteins
    coupled to CNBr-Sepharose 4B
    and purified by human RBC protein-AA-Sepharose 4B

  Antihemocyanin-AA and antimyoglobin-AA

Our laboratory was the first to use ion-exchange HPLC to detect the formation of hemoglobin–AA in vivo.[36,37] Because the fast-moving hemoglobin HPLC profiles in mice and humans resemble each other closely on the Synchropak CM 300 cation-exchange column, we took 32 C57/BL mice and divided them equally into an alcohol-fed group and a control group and then pairfed them by liquid diets. After 3 wk, peaks 3', 4, and 11 of RBC lysates from the alcohol-fed mice constituted 0.059 ± 0.032 ($\pm$SD), $0.174 \pm .002$, and $2.873 \pm 0.057\%$ of the total hemoglobin; these values were significantly higher ($p < 0.001$) than the respective peaks in RBC lysates obtained from the control mice, [0.055 ± 0.008, 0, and $2.529 \pm 0.066\%$ (Fig. 3)]. Although this HPLC system clearly distinguished alcohol-fed mice from pair-fed controls, it failed to diagnose alcohol abuse reliably in a subsequent clinical study that included 10 alcoholics and 10 control subjects.

Our data were in agreement with the negative results of several reports, including those published by Homaiden et al. (1984), Gordis and Herschkopt (1986), and Stockham and Blanke (1988) (Table 5, *see* page 174).[54-56] It should be noted that all the negative reports cited in Table 5 came from laboratories that used highly sensitive chromatographic techniques. By contrast, the two positive reports, Stevens et al. (1981) and Hoberman and Chiodo (1982), relied on chromatographic approaches known to exhibit much lower resolution power and lower sensitivity.[33,53] Thus, it seems that chromatography alone is probably inadequate to directly (without further chemical treatment) measure hemoglobin–AAs in erythrocytes obtained from alcoholic patients.

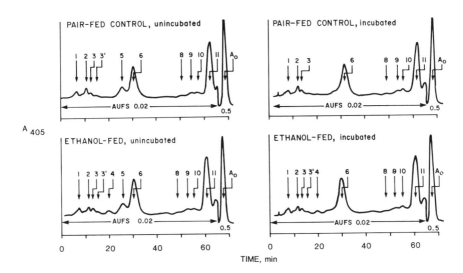

Fig. 3. Increased level of fast-moving hemoglobin fractions, i.e., peaks 3', 4 and 11, in lysates obtained from ethanol-fed mice when compared to pair-fed controls. Erythrocytes from ethanol- and pair-fed control mice were studied without (panels on the left) and with (panels on the right) prior incubation in buffered 154 mM NaCl (pH 7.4) for 5 h at 37°C. With prior incubation, the labile adducts between glucose and hemoglobin dissociated; and accordingly, peak 5 disappeared, and peak 11 also decreased in peak height.

Realizing the difficulties with any direct method of detecting hemoglobin, or plasma protein-associated AA, Peterson and Polizzi in 1987 developed a sensitive fluorogenic HPLC method that entailed the reaction of acetaldehyde with two molecules of 1,3-cyclo-hexanedione in the presence of ammonium ion to form a fluorescent species that can be identified by reverse-phase HPLC and by fluoro-metric detector.[46] The assay was found to be specific and sensitive in the picomole range.[57,58] It also exhibited intrassay precision of < 3.5% and interassay precision less than 15%. Although these investigators claimed their method measures only acetaldehyde that is reversibly bound to plasma proteins and hemoglobin, it is likely that this may not be the case. Recent studies by Baraona and his colleagues indicate that there is increased reversible binding of acetaldehyde to erythrocytes

Table 5
Fast-Moving Hemoglobin Fractions in Alcoholics vs Controls

Reference	Methods	\multicolumn{2}{c}{Alcoholics}	\multicolumn{2}{c}{Controls}	p
		n	Results	n	Results	
33	Chromatography	10	$12.1 \pm 2.8\%$		$8.8 \pm 0.8\%$	< 0.05
53	Chromatography	43	$7.5 \pm 0.9\%$	41	$6.2 \pm 0.6\%$	< 0.001
54	Chromatography	16	$6.0 \pm 0.9\%$	7	5–8%	NS
55	IEF	70	Anodal CNBH band	15	Anodal CNBH band	NS
56	Affinity column	31	$4.4 \pm 0.9\%$	28	$4.3 \pm 0.8\%$	NS

in alcoholics and that much of this binding can be explained by the formation of thiazolidine with cysteine.[59,60] Cysteine content in erythrocytes has been reported to double in alcoholics when compared to controls. Additionally, the levels of cysteine and reversibly bound acetaldehyde in erythrocytes can remain high for up to 2 wk after withdrawal of alcohol. Based on these observations, it is highly probable that the fluorogenic HPLC method developed by Peterson and Pollizi measures mainly reversible AAs, bound to hemoglobin and plasma proteins as well as to small molecules such as cysteine.

Peterson and his colleagues have obtained interesting and important data with their fluorogenic HPLC method for measuring AAs in venous blood.[46,61–63] They found that venous blood even from teetotalers contained plasma-associated ($0.43 \pm 0.04\ \mu M$) and hemolysate-associated acetaldehyde ($1.80 \pm 0.38\ \mu M$), suggesting that endogenous acetaldehyde is present in humans.[46] By comparison, alcoholic patients exhibited significantly higher plasma-associated ($0.77 \pm 0.31\ \mu M$) and hemolysate-associated ($3.75 \pm 2.72\ \mu M$) acetaldehyde levels than those found in teetotalers.

In order to understand the kinetics of formation and degradation of plasma and hemolysate-associated acetaldehyde, Peterson et al. (1988) acutely administered to volunteers (social drinkers) 0.3 g/kg of ethanol in a clinical research center setting and then collected venous blood at frequent intervals.[61] These investigators found that peak levels of plasma- and hemolysate-associated acetaldehyde were attained in about 1.5 h. Following this, both plasma- and hemolysate-associated acetaldehyde returned to baseline within 3.5 h. When compared with the levels found in teetotalers, plasma-associated acetaldehyde of volunteers given ethanol returned to within 1 SD of levels found in teetotalers by 5 d, whereas hemolysate-associated acetaldehyde remained elevated (compared to teetotalers) for up to 28 d.

Peterson et al. (1990) also measured plasma-, hemolysate-, platelet-, and lymphocyte-associated acetaldehyde in miniature swines maintained on 0, 2, or 6 g of ethanol/kg/d for 8 months ($n = 6$ for each group).[63] Measurements of hemolysate-associated acetaldehyde (250 ± 47 nmol/g vs 203 ± 33, $p < 0.05$) and platelet-associated acetaldehyde (0.45 ± 0.34 nmol/3×10^8 platelets vs 0.15 ± 0.16, $p = 0.05$) were found to be useful in discriminating drinking pigs (pigs on 2 and 6 g of ethanol/kg/d combined) from control pigs. Analyses of plasma- and lymphocyte-associated acetaldehyde were not useful as a marker of alcohol consumption. Increased amounts of bound acetaldehyde were found in several blood compartments 8 h after cessation of ethanol ingestion. Based on these results, it seems clear that plasma- and hemolysate-associated acetaldehyde represent reversibly bound acetaldehyde and that they are much less stable than the acetaldehyde found in protein–AAs, such as the 37 kDa liver protein–AA.

Our laboratory and that of Niemela and his co-workers have developed antibodies that can specifically recognize AAs as epitopes including those in hemoglobin–AA.[47,48,64] With antihemocyanin-AA IgG as the antibody and using a sandwich enzyme-linked immunosorbent assay (ELISA), we have examined the extent of hemoglobin/ AA formation in miniature swine fed varying amounts of ethanol, i.e., 0, 2, and 6 g of ethanol/kg/d. Figure 4 depicts the ELISA results before initiation of feeding and at the 8-mo time point. There was no change in OD_{405} in the control group (group III) by the eight months; however, the OD_{405} from groups I and II increased significantly. The largest increment was observed in the hemoglobin of miniature pigs fed 6 g of ethanol/kg/d, i.e., group I.

Using antibovine serum albumin–AA cross-absorbed with human RBC proteins coupled to CNBr-Sepharose 4B and then purified by human RBC protein–AA-Sepharose 4B, Niemela and Israel (1991) have recently reported in an abstract their experience with the use of hemoglobin–AA as a diagnostic marker of alcohol abuse in humans.[48] They measured hemoglobin–AA from erythrocytes of 69 alcohol abusers, 67 social drinkers, 18 abstainers, and 76 hospitalized patients.

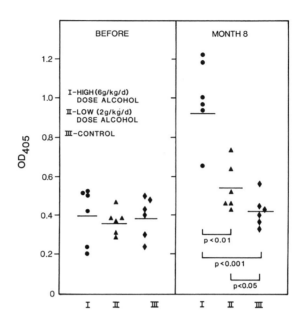

Fig. 4. Formation of hemoglobin–AA in miniature pigs fed alcohol chronically for 8 mo. Group I: pigs that were fed high-dose alcohol, 6 g/kg/d; group II: pigs that were fed low-dose alcohol, 2 g/kg/d; and Group III: control pigs.

They found that hemoglobin–AA values were highest in the alcohol abusers ($p < 0.01$) but were also elevated in social drinkers ($p < 0.05$) when compared with the abstainers. Eighty-three of the 169 alcohol abusers (40%) and 17 of the 67 social drinkers (25%) exhibited abnormally elevated hemoglobin–AA values. By direct comparison, they found that 62% of alcoholics exhibited abnormally elevated GGT activities and 40% increased MCV. Upon abstinence from alcohol, hemoglobin–AA levels in the alcoholics decreased as a function of time over 1–3 wk. These preliminary data are encouraging but certainly fall short of expectations.

Clearly, much more research is needed to improve the methods to detect hemoglobin–AAs in alcoholics. Proper storage of samples remains a problem. The choice of method to prepare antibodies to detect hemoglobin–AAs is a critical issue. The choice of proper test format will also require further research.

Measurement
of Serum Protein-AAs In Vivo

Our laboratory has used both antihemocyanin–AA IgG and antimyoglobin–AA IgG in a two-site (or sandwich) ELISA assay system to measure serum protein–AAs in alcoholics and teetotalers.[47] The intra- and interassay coefficients of variance were found to be about 8%. Using sandwich ELISA and antihemocyanin–AA IgG, the OD$_{405}$ readings for sera of control subjects and alcoholic patients were 0.036 ± 0.033 (±SEM; $n = 28$) and 0.150 ± 0.088 ($n = 28$), respectively, and the sensitivity was 64%. It was found that serum protein-AAs reacted more strongly with antimyoglobin–AA IgG than with antihemocyanin–AA IgG. Accordingly, the sensitivity increased to 71% when antimyoglobin–AA IgG was used in the ELISA assay. By direct comparison, the sensitivities of GGT, ALT, MCV, and AST were lower than serum protein–AA ELISA, i.e., 57, 40, 36, and 30%, respectively. By immunotransblot, two serum protein-AAs were visualized on nitrocellulose paper, and they corresponded to molecular weights of 103 kDa and 50 kDa. The 50 kDa protein-AA did not comigrate with albumin since the known molecular weight of albumin is about 64 kDa.

Conclusions

There is now ample evidence that both stable and unstable protein-AAs, including serum protein–AAs and hemoglobin–AA, do form in vivo during chronic exposure to alcohol. Unstable protein-AAs and AAs involving small molecules, e.g., cysteine, can be detected by fluorogenic HPLC.[46] To date, clinical studies indicate that stable protein–AAs in blood circulation are present in very low concentrations and that they cannot be detected by direct chromatography or by IEF.[54,55] However, by sensitive immunological methods, at least two laboratories have been successful recently in detecting the formation of either stable serum protein–AAs or stable hemoglobin–AA in heavy drinkers and alcoholics.[47,48] The published sensitivities of ELISA assays for stable serum protein–AAs and hemoglobin–AA are about the same as γ-glutamyl transferase (GGT) or slightly better. Improvement in sample preservation, development of stronger and more specific antibodies, introduction of improved test formats, and conducting larger clinical studies will be needed to better define the true sensitivities and specificities of stable serum protein–AAs and hemoglobin–AA measurements as markers of alcohol abuse.

Acknowledgments

This work is supported by grants from the National Institute on Alcohol Abuse and Alcoholism (R01 07647) and US Department of Veterans Affairs Merit Review Funds.

References

[1]Y. Israel (1985) Markers of alcohol consumption and task ahead, in Early Identification of Alcohol Abuse, Research Monograph No. 17, N. C. Chang and H. M. Chao, eds. (PHHS Publication, Rockville, MD), pp. xi–xv.

[2]A. W. K. Chan (1990) Biochemical markers for alcoholism, in Children of Alcoholics: Critical perspectives. M. Windle and J. S. Searles, eds. (Guilford, New York), pp. 39–71.

[3]M. L. Selzer (1972) The Michigan Alcoholism Screening Test: The quest for a new diagnostic instrument. Am. J. Psychol. 127, 1653–1658.

[4]J. A. Ewing (1984) Detecting alcoholism: The CAGE questionnaire. JAMA 252, 1905–1907.

[5]M. W. Bernadt, C. Taylor, and J. Mumford (1982) Comparison of questionnaire and laboratory tests in the detection of excessive drinking and alcoholism. Lancet 1, 325–328.

[6]T.-K. Li (1977) Enzymology of human alcohol metabolism. Adv. Enzymol. 45, 427–483

[7]C. S. Lieber and L. M. DeCarli (1970) Hepatic microsomal ethanol-oxidizing system. In vitro characteristics and adaptive properties in vivo. J. Biol. Chem. 245, 2505–2512.

[8]H. W. Goedde and D. P. Agarwal (1989) Acetaldehyde metabolism: genetic variation and physiological implications, Alcoholism: Biomedical and Genetic Aspects. H. W. Goedde and D. P. Agarwal, eds. (Pergamon, New York), pp. 21–56.

[9]C. S. Lieber and L. M. DeCarli (1972) The role of the hepatic microsomal ethanol oxidizing system (MEOS) for ethanol metabolism in vivo. J. Pharmacol. Exp. Ther. 181, 279–287.

[10]C. J. P. Eriksson (1980) Problems and pitfalls in acetaldehyde determinations. Alcoholism: Clin. Exp. Res. 4, 22–29.

[11]C. J. P. Eriksson (1983) Human blood acetaldehyde concentration during ethanol oxidation. Pharmacol. Biochem. Behav. 18, 141–150.

[12]K. O. Lindros (1983) Human blood acetaldehyde levels: with improved methods, a clearer picture emerges. Alcoholism: Clin. Exp. Res. 7, 70–75.

[13]S.-J. Yin and T.-K. Li (1989) Genetic polymorphism and properties of human alcohol and aldehyde dehydrogenase: Implications for ethanol metabolism and toxicity, *Molecular Mechanisms of Alcohol: Neurobiol. Metabolism*. G. Y. Sun, P. K. Rudeen, W. G. Wood, Y. H. Wei and A. Y. Sun, eds. (Humana Press, Clifton, NJ), pp. 227–248.

[14]K. O. Lindros, A. Stowell, P. Pikkarainen, and M. Salaspuro (1980) Elevated blood acetaldehyde in alcoholics with accelerated ethanol elimination. *Pharmacol. Biochem. Behav.* **13**, 119–124.

[15]K. Matthewson, A. L. H. Mardini, K. Barlett, and C. O. Record (1986) Impaired acetaldehyde metabolism in patients with non-alcoholic liver disorders. *Gut* **27**, 756–764.

[16]H. R. Thomasson, T.-K. Li, and D. W. Crabb (1990) Correlations between alcohol-induced flushing, genotypes for alcohol and aldehyde dehydrogenase and alcohol elimination rates. *Hepatology* **12**, 264.

[17]E. Mezey, A. L. Imbembo, J. J. Potter, and P. Holt (1975) Ethanol production and hepatic disease following jejunoileal bypass for morbid obesity. *Am. J. Clin. Nutr.* **28**, 1277–1283.

[18]E. Baraona, R. Julkunen, L. Tannenbaum, et al. (1986) Role of intestinal bacterial overgrowth in ethanol production and metabolism in rats. *Gastroenterology* **90**, 103–110.

[19]I. R. McManus, E. Brotsky, and R. E. Olson (1966) The origin of ethanol in mammalian tissues. *Biochim. Biophys. Acta* **121**, 167–170.

[20]L. I. Malkin and D. M. Greenberg (1964) Purification and properties of threonine or allothreonine aldolase from rat liver. *Biochim. Biophys. Acta* **85**, 117–131.

[21]F. J. Lionetti, N. L. Fortier, and J. A. Jedziniak (1964) Acetaldehyde, a product of deoxynucleoside metabolism in human erythrocyte ghosts. *Proc. Soc. Exp. Bio. Med.* **116**, 1080–1082.

[22]H. L. Fleshood and H. C. Pitot (1970) The metabolism of O-phosphorylethanolamine in animal tissues. *J. Biol. Chem.* **245**, 4414–4420.

[23]R. C. San George and H. D. Hoberman (1986) Reaction of acetaldehyde with hemoglobin. *J. Biol. Chem.* **261**, 6811–6821.

[24]L. Lumeng and P. J. Durant (1985) Regulation of the formation of stable adducts between acetaldehyde and blood proteins. *Alcohol* **2**, 397–400.

[25]T. Donohue D. Tuma, and M. Sorrell (1983) Acetaldehyde adducts with proteins: Binding of (^{14}C) acetaldehyde to serum albumin. *Arch. Biochem. Biophys.* **220**, 239–246.

[26]K. Gaines, J. Salhary, D. Tuma, and M. Sorrell (1977) Reactions of acetaldehyde with human erythrocyte membrane proteins. *FEBS Letters* **75**, 115–119.

[27]F. Nomura and C. S. Lieber (1981) Binding of acetaldehyde to rat liver microsomes: Enhancement after chronic alcohol consumption. *Biochem. Biophys. Res. Comm.* **100**, 131–137.

[28]D. Tuma, R. Jennett, and M. Sorrell (1987) The interaction of acetaldehyde with tubulin. *Ann. NY Acad. Sci.* **492,** 277–286.

[29]D. S. Xu, R. B. Jennett, S. L. Smith, M. F. Sorrell, and D. J. Tuma (1989) Covalent interactions of acetaldehyde with the actin/microfilament system. *Alcohol Alcoholism* **24,** 281–289.

[30]T. Mauch, D. Tuma, and M. Sorrell (1987) The binding of acetaldehyde to the active site of ribonuclease: Alterations in catalytic activity and effects of phosphate. *Alcohol* **22,** 103–112.

[31]D. J. Tuma, M. R. Newman, T. M. Donohue, and M. F. Sorrell (1987) Covalent binding of acetaldehyde to proteins: Participation of lysine residues. *Alcoholism: Clin. Exp. Res.* **11,** 579–584.

[32]T. J. Mauch, T. M. Donohue, R. K. Zetterman, and M. F. Sorrell (1985) Covalent binding of acetaldehyde to lysine-dependent enzymes can inhibit catalytic activity. *Hepatology* **5,** 1056.

[33]V. J. Steven, U. J. Fantl, C. B. Newman, R. V. Sims, A. Cerami, and C. M. Peterson (1981) Acetaldehyde adducts with hemoglobin. *J. Clin. Invest.* **67,** 361–369.

[34]L. Lumeng, R. Minter, and T.-K. Li (1982) Distribution of stable acetaldehyde adducts in blood under physiological conditions. *Fed. Proc.* **41,** 765.

[35]L. B. Nguyen and C. M. Peterson (1986) Differential modification of hemoglobin chains by acetaldehyde. *Proc. Soc. Exp. Bio. Med.* **181,** 151–156.

[36]L. Lumeng and R. G. Minter (1985) Formation of acetaldehyde-hemoglobin adducts in vitro and in vivo demonstrated by high performance liquid chromatograph. *Alcoholism: Clin. Exp. Res.* **9,** 209.

[37]L. Lumeng and R. Minter (1985) Formation of acetaldehyde-hemoglobin adducts in vitro and during chronic alcohol ingestion. *Clin. Res.* **33,** 529A.

[38]P. Sillanaukee and T. Koivula (1990) Detection of a new acetaldehyde-induced hemoglobin fraction Hb$_{A1ach}$ by cation exchange liquid chromatography. *Alcoholism: Clin. Exp. Res.* **14,** 842.

[39]S. M. Gapstur, E. G. DeMaster, J. D. Belcher, J. D. Potter, M. D. Gross (1991) The formation of stable hemoglobin adducts in human red blood cells exposed to ethanol and acetaldehyde. *Alcoholism: Clin. Exp. Res.* **15,** 378.

[40]R. C. Lin, R. S. Smith, and L. Lumeng (1988) Detection of a protein-acetaldehyde adduct in the liver of rats fed alcohol chronically. *J. Clin. Invest.* **81,** 615–619.

[41]U. J. Behrens, M. Hoerner, J. M. Lasker, and C. S. Lieber (1988) Formation of acetaldehyde adducts with ethanol-inducible P450IIE1 in vivo. *Biochem. Biophys. Res. Comm.* **154,** 584–590.

[42]U. J. Behrens, X.-L. Ma, E. Baraona, and C. S. Lieber (1989) Acetaldehyde-collagen adducts in CCl_4-induced liver injury in rats. *Hepatology* **10,** 608.

[43]Y. Israel, E. Hurwitz, O. Niemela, and R. Arnon (1986) Monoclonal and polyclonal antibodies against acetaldehyde-containing epitopes in acetaldehyde-protein adducts. *Proc. Nat. Acad. Sci.* **83,** 7923–7927.

[44]M. Hoerner, U. J. Behrens, T. Worner, and C. S. Lieber (1986) Humoral immune response to acetaldehyde adducts in alcohol patients. *Res. Comm. Chem. Pathol. Pharmacol.* **54**, 3–12.

[45]O. Niemela, F. Klajner, H. Orrego, E. Vidins, L. Blendis, and Y. Israel (1987) Antibodies against acetaldehyde-modified protein epitopes in human alcoholics. *Hepatology* **7**, 1210–1214.

[46]C. M. Peterson and C. M. Polizzi (1987) Improved method for acetaldehyde in plasma and hemoglobin-associated acetaldehyde. *Alcohol* **4**, 477–480.

[47]R. C. Lin, L. Lumeng, S. Shahidi, T. Kelly, and D. Pound (1990) Protein-acetaldehyde adducts in serum of alcoholic patients. *Alcoholism: Clin. Exp. Res.* **14**, 438–443.

[48]O. Niemela and Y. Israel (1991) Evaluation of acetaldehyde-hemoglobin adducts as markers of alcohol consumption in humans. *Gastroenterology* **100**, A780.

[49]R. C. Lin and L. Lumeng (1989) Further studies on the 37KD liver protein-acetaldehyde adduct that forms in vivo during chronic alcohol ingestion. *Hepatology* **10**, 807–814.

[50]R. C. Lin and L. Lumeng (1990) Formation of the 37-KD protein-acetaldehyde adduct in liver during alcohol treatment is dependent on alcohol dehydrogenase activity. *Alcoholism: Clin. Exp. Res.* **14**, 766–770.

[51]R. C. Lin, M. Fillenwarth, R. Minter, and L. Lumeng (1990) Formation of the 37–KD protein-acetaldehyde adduct in primary cultured rat hepatocytes exposed to alcohol in vitro. *Hepatology* **11**, 401–407.

[52]L. Lumeng and R. C. Lin (1991) Formation of a 37 kilodalton liver protein-acetaldehyde adduct in vivo and in liver cell culture during chronic alcohol exposure. *Ann. N.Y. Acad. Sci.* **625**, 793–801.

[53]H. D. Hoberman and S. M. Chiodo (1982) Elevation of the hemoglobin Al fraction in alcoholism. *Alcoholism: Clin. Exp. Res.* **6**, 260–266.

[54]F. R. Homaidan, L. J. Kricka, P. M. S. Clark, S. R. Jones, and T. P. Whitehead (1984) Acetaldehyde-hemoglobin adducts: An unreliable marker of alcohol abuse. *Clin. Chem.* **30**, 480–482.

[55]E. Gordis and S. Herschkopf (1986) Application of isoelectric focusing in immobilized pH gradients to the study of acetaldehyde-modified hemoglobin. *Alcoholism: Clin. Exp. Res.* **10**, 311–319.

[56]T. I., Stockham and R. V. Blanke (1988) Investigation of an acetaldehyde-hemoglobin adduct in alcoholics. *Alcoholism: Clin. Exp. Res.* **12**, 748–754.

[57]W. L. Stahovec and K. Mopper (1984) Trace analysis of aldehydes by phase high-performance liquid chromatograph. *J. Chromatogr.* **298**, 399–406.

[58]N. S. Ung-Chhun and M. A. Collins (1987) Estimation of blood acetaldehyde during ethanol metabolism: a sensitive HPLC/Fluorescence microassay with negligible artifactual interference. *Alcohol* **4**, 473–476.

[59]E. Baraona, C. Di Padova, J. Tabasco, and C. S. Lieber (1987) Red blood cells: A new major modality for acetaldehyde transport from liver to other tissues. *Life Sci.* **40**, 253–258.

[60]R. Hernandez-Munoz, E. Baraona, I. Blacksberg, and C. S. Lieber (1989) Characterization of the increased binding of acetaldehyde to red blood cells in alcoholics. *Alcoholism: Clin. Exp. Res.* **13**, 654–659.

[61]C. M. Peterson, L. Jovanovic-Peterson, and F. Schmid-Formby (1988) Rapid association of acetaldehyde with hemoglobin in human volunteers after low dose ethanol. *Alcohol* **5**, 371–374.

[62]C. M. Peterson and B. K. Scott (1989) Studies of whole blood associated acetaldehyde as a marker for alcohol intake in mice. *Alcoholism: Clin. Exp. Res.* **13**, 845–848.

[63]C. M. Peterson, B. K. Scott, G. Y. Sun, and A. Y. C. Sun (1990) A comparative blinded study in miniature swine of whole blood-, hemoglobin-, platelet-, plasma-, and lymphocyte-associated acetaldehyde as markers for ethanol intake. *Alcoholism: Clin. Exp. Res.* **14**, 717–720.

[64]O. Niemela, Y. Israel, Y. Mizoi, T. Fukunaga, and C. J. P. Eriksson (1990) Hemoglobin-acetaldehyde adducts in human volunteers following acute ethanol ingestion. *Alcoholism: Clin. Exp. Res.* **14**, 838–841.

[65]R. K. Pullarkat and S. Raguthu (1985) Elevated urinary dolichol levels in chronic alcoholics. *Alcoholism: Clin. Exp. Res.* **9**, 28–30.

[66]U.-M. Korri, H. Nuutinen, and M. Salaspuro (1985) Increased blood acetate: A new laboratory marker of alcoholism and heavy drinking. *Alcoholism: Clin. Exp. Res.* **9**, 468–471.

[67]D. D. Rutstein, R. L. Veech, R. J. Nickerson, M. E. Felyer, A. A. Vernon, L. L. Needham, P. Kishore, and S. B. Thacker (1983) 2,3-Butanediol: An unusual metabolite in the serum of severely alcoholic men during acute intoxication. *Lancet* **2**, 534–537.

[68]B. K. Tang, P. Devenyi, D. Teller, and Y. Israel (1986) Detection of an alcohol specific product in urine of alcoholics. *Biochem. Biophys. Res. Comm.* **140**, 924–927.

[69]B. Nalpas, A. Vassault, A. LeGuillou, B. Lesgourgues, N. Ferry, B. Lacour, and P. Berthelot (1984) Serum activity of mitochondrial aspartate aminotransferase: A sensitive marker of alcoholism with or without alcoholic hepatitis. *Hepatology* **4**, 893–896.

[70]C. C. Lin, J. J. Potter, and E. Mezey (1984) Erythrocyte aldehyde dehydrogenase activity in alcoholism. *Alcoholism: Clin. Exp. Res.* **8**, 539–541.

[71]P. Puchois, M. Fontan, J. L. Gentilini, P. Gelez, and J. C. Fruchart (1984) Serum apolipoprotein A-II, A biochemical indicator of alcohol abuse. *Clin. Chim. Acta* **185**, 185–189.

[72]H. Stibler, S. Borg, and C. Allgulander (1980) Abnormal microheterogeneity of transferrin: A new marker of alcoholism? *Substance Alcohol Actions/Misuse* **1**, 247–252.

Measuring Alcohol Consumption by Transdermal Dosimetry

Michael Phillips

Introduction

Physiological Basis of the TDD

More than a century ago, Anstie found that alcohol is excreted through the skin into the sweat.[1] This pathway is quantitatively insignificant compared to hepatic metabolism, yet it provides a window on ethanol metabolism in vivo. The transdermal dosimeter (TDD) is a device that allows an observer to spy on ethanol metabolism in the body by peering through the cutaneous window.

Since ethanol is distributed fairly equally throughout the entire body, the concentration of ethanol in the sweat at any time is usually very close to the concentration in the blood.[2,3] The physiologic basis of the TDD rests on two assumptions: that the concentrations of ethanol in blood and sweat are similar and that the TDD can collect fluid from the surface of the skin at a constant rate for a defined period of time. If these two conditions are fulfilled, the concentration of ethanol in the collected fluid specimen should be the same as the mean concentration of ethanol in the blood during the period of collection. The advantage of the TDD is that it can collect a fluid sample in a noninvasive fashion over a period of days and provide two kinds of information:

From: *Measuring Alcohol Consumption*
Eds.: R. Litten and J. Allen ©1992 The Humana Press Inc.

1. Qualitative—did the wearer of the TDD drink any alcohol during the sampling period? The presence of alcohol in the TDD is direct evidence that some alcohol was consumed while the TDD was worn.
2. Quantitative— the amount of ethanol in the TDD allows some inferences to be made about the quantity of alcohol that was consumed.

Technical Problems
in Development of the TDD

Although the principle of the TDD is simply stated, development of a clinically useful device has been accompanied by some formidable technical requirements and difficulties:

1. Adhesiveness—the TDD must remain attached to the skin for up to 8 d on normal subjects leading an active life outside a hospital;
2. Uptake linearity—fluid must be collected from the surface of the skin continuously at a steady rate during the entire collection period;
3. Acceptability to wearer—the TDD should be comfortable and acceptable to the wearer;
4. Acceptability to user—the device should be "user friendly", i.e., easy to apply, remove, assay, and interpret;
5. Sample retention—ethanol in the collected fluid must not be lost by leakage or back-diffusion across the skin; and
6. Calibration—there should be a known relationship between the amount of ethanol in the TDD and the amount of ethanol the wearer has ingested.

Evolution of the TDD

The present TDD is a fourth-generation device. The evolution of the first three generations has been recently reviewed.[4] In summary, the first-generation TDD was a hand-assembled device made of pads impregnated with sodium chloride. Studies with this TDD demonstrated that it could collect fluid from the surface of the skin in a linear fashion for up to 1 wk; also, ethanol could be detected in the TDD after it was worn during supervised consumption of alcohol.[5,6] The second-generation TDD was smaller, more comfortable, and constructed in a more reproducible fashion. It had improved linearity of uptake over an 8-d period,[7] and when worn by drinking volunteers, the concentration of ethanol in the TDD varied exponentially with the average daily dosage of ethanol and linearly with the mean blood alcohol concentration during the collection period.[7,8] Field studies with this TDD demonstrated that it was well-accepted by normal subjects and could detect alcohol consumption by subjects who claimed to have drank no alcohol at all.[9,10] The third-generation TDD was manufactured by 3M and incorporated Silicalyte[R] as a binding agent for ethanol to limit losses from back-diffusion across the skin. An unpublished inhouse study performed by 3M confirmed the findings of the second-generation TDD.

The Fourth-Generation TDD

This device incorporates a number of evolutionary improvements, particularly in ease of use and assay. The absorptive pad comprises absorbent polyester laminated with cloth impregnated with activated carbon, which captures ethanol in the sample by adsorptive binding, thereby limiting losses from leakage or back-diffusion across the skin. Assay of ethanol in the sample may be performed as an office procedure by heating the collecting pad in a sealed container and assaying the head-space ethanol with a hand-held electrochemical detector.[11] This TDD is currently being evaluated in an outpatient calibration study.

Conclusions

Accuracy

Two inpatient drinking studies have demonstrated a high degree of accuracy: Ethanol in the TDD varied linearly with mean blood–alcohol concentration during the period the TDD was worn, as well as linearly with the square root of average daily dosage of ethanol (in g/kg/d).[6,8]

Ease of Use and Cost

The TDD is easy to use and can be applied, removed, and assayed by nonmedical and nonscientific personnel after a short period of training. Cost has not yet been determined but should compare favorably with routine laboratory blood assays for liver enzymes.

Utility for Various Settings

The TDD is used most appropriately in an outpatient setting where patients are seen on a weekly basis.

Latency and Retrospective
Accuracy Capabilities

The TDD is effective only for the period it is worn. It gives information about alcohol consumption between the times of application and removal.

Acceptability to Practitioner and Patients

The TDD is acceptable to practitioners since it is noninvasive, highly sensitive, and specific for alcohol consumption. An outpatient study of volunteers wearing the TDD demonstrated a high degree of acceptability: Most patients forgot about its presence within a few hours and described it as similar to wearing a bandaid.[9]

Unique Benefits of TDD

1. High sensitivity—early versions detected intake of alcohol at level of 0.5 g/kg/d. More recent versions have higher sensitivity, so that the probability of a false-negative is very low.
2. High specificity—the assay reacts only to ethanol, and does not yield false-positives.
3. Clearly defined sampling period—the TDD provides information about alcohol consumption between the times of application and removal.

Problems with Measurement

The major problem is susceptibility to tampering. There is probably no way to prevent a patient from tampering with the TDD. However, it is possible to build in a sensitive detector to determine if tampering has occurred. A number of methods are currently being evaluated. One method that looks especially promising is a new security tape that changes in appearance when detached and remains changed even when readhered.

Needs for Further R&D

The major needs are calibration of the most-recent version of the TDD, to determine its dose-response characteristics, and development of a foolproof method for the detection of tampering.

References

[1]F. D. Anstie (1874) Final experiments on the elimination of alcohol from the body. *Practitioner* **13,** 15.

[2]G. L. S. Pawan and K. Grice (1968) Distribution of alcohol in urine and sweat after drinking. *Lancet* **2,** 1016.

[3]S. W. Brusilow and E. H. Gordes (1966) The permeability of the sweat gland to non-electrolytes. *Am. J. Dis. Child* **112,** 328.

[4]M. Phillips (1989) Measurement of alcohol consumption by transdermal dosimetry, *Diagnosis of Alcohol Abuse.* R. R. Watson, ed. (CRC Press, Boca Raton, FL), pp. 207–216.

[5]M. Phillips, R. E. Vandervoort, and C. E. Becker (1977) Long-term sweat collections using salt-impregnated pads. *J. Invest. Dermatol.* **68** (4), 221–224.

[6]M. Phillips, R. E. Vandervoort, and C. E. Becker (1978) A sweat test for alcohol consumption. *Currents in Alcohol.* **3,** 505–514.

[7]M. Phillips (1980) An improved adhesive patch for long-term collection of sweat. *Biomater. Med. Devices Art. Organs* **8** (1), 13–21.

[8]M. Phillips and M. H. McAloon (1980) A sweat-patch test for alcohol consumption: Evaluation in continuous and episodic drinkers. *Alcoholism: Clin. Exp. Res.* **4** (4), 391–395.

[9]M. Phillips (1984) Subjective responses to the sweat-patch test for alcohol consumption. *Adv. Alcohol Subst. Abuse* **3** (4), 61–67.

[10]M. Phillips (1984) Sweat-patch testing detects accurate self-reports for alcohol consumption. *Alcoholism: Clin. Exp. Res.* **8** (1), 51–53.

[11]M. Phillips (1982) Sweat-patch test for alcohol consumption: Rapid assay with an electrochemical detector. *Alcoholism: Clin. Exp. Res.* **6** (4), 532–534.

Assessment of Ethanol Consumption with a Wearable, Electronic Ethanol Sensor/Recorder

Robert M. Swift and Larry Swette

Introduction

Accurate and reliable measurement of alcohol consumption over extended time periods is important for treatment, research, and forensic applications. In a clinical setting, assessing alcohol consumption is important for identifying problem drinking and monitoring treatment compliance in individuals undergoing substance-abuse treatment. In a research setting, determining consumption is necessary to assess treatment outcomes and to understand patterns and amounts of drinking behaviors in various cohorts. Forensic applications include monitoring alcohol use in special populations, including transportation workers, drunk drivers, and impaired professionals.

The ideal method for measuring alcohol consumption should be widely applicable, transportable to multiple settings across a range of users, inexpensive, accurate and valid, highly specific and sensitive for ethanol, and easily accepted by patient and clinician or subject and researcher.[1] A variety of methodologies has been developed to gather information regarding alcohol consumption. These include verbal report measures,[2-7] measurement of biochemical markers associated

From: *Measuring Alcohol Consumption*
Eds.: R. Litten and J. Allen ©1992 The Humana Press Inc.

with alcohol use,[8–11] and direct measurement of urine, blood, saliva, breath, or sweat alcohol levels. Although each of these methodologies has advantages and disadvantages, no single method currently exists that meets all of the above criteria to easily and reliably quantify patterns of alcohol consumption over a period of days or weeks.

One promising method of monitoring alcohol use is through detection of ethanol in cutaneous perspiration. The clinical pharmacology of transcutaneous ethanol is poorly understood, but it has long been known that ingested ethanol is partitioned into the body water and that measurable quantities of ingested ethanol are excreted through the human skin.[12,13] Measurements of human sweat alcohol concentrations in liquid and gas phases find approximately equal concentrations in blood and other body fluids, when corrected for evaporation and water content.[12–15] Transcutaneous ethanol appears to be excreted by exocrine sweat glands and to passively diffuse across the skin.[14,16]

The alcohol dosimeter, or "sweat patch," was designed to be a portable, wearable, noninvasive device that utilizes ethanol in perspiration to yield qualitative, cumulative data on ethanol consumption over a 7- to 10-day period.[17–19] However, field trials of the sweat patch method elicited a number of problems that limit its utility. These include problems with storage, extraction, and measurement of ethanol and losses caused by evaporation, leakage, back-diffusion, and bacterial metabolism of ethanol.[20,21] In addition, alcohol dosimeter measurement of ethanol consumption is cumulative and does not give information about the timing and pattern of alcohol use.

Recent advances in electrochemical detection technology have led to the creation of shorted electrochemical cells that oxidize ethanol and generate an electrical current proportional to the concentration of ethanol vapor.[22] This type of cell has been adapted for the measurement of ethanol concentration in expired air and is the analytic component in commercially available, portable, battery-operated breathalyzers. These instruments have a high correlation with blood alcohol concentration (BAC) and are commonly used by clinicians and law-enforcement officials to determine recent ethanol use.[23,24] However, breathalyzer technology is not suitable for continuous long-term measurement of alcohol and is not passive, requiring active subject participation.

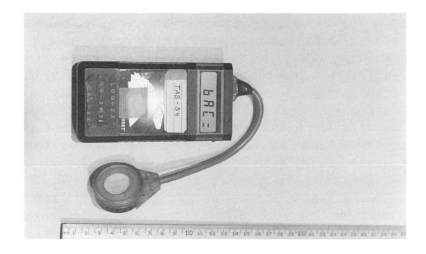

Fig. 1. Photograph of the Giner, Inc. Transdermal Ethanol Sensor/ Recorder (TAS).

The adaption of electrochemical detection of ethanol to the measurement of transdermal ethanol seems an ideal approach for noninvasive, long-term monitoring of ethanol use. By modifying an electrochemical detector to provide continuous rather than episodic monitoring of transcutaneous ethanol, accurate quantitative and temporal tracking of ethanol consumption over extended periods may be determined.

Transdermal Alcohol Sensor/Recorder

The Transdermal Alcohol Sensor/Recorder (TAS), a wearable, battery-operated device, has been developed by Giner, Inc. TAS continuously samples transdermal ethanol vapor and stores digital data at intervals for later analysis. A photograph of an advanced prototype is shown in Fig. 1. Unlike the alcohol dosimeter, TAS provides real time rather than cumulative monitoring of alcohol use. TAS consists of two components: an electrochemical sensor that detects ethanol vapor at the surface of the skin and a data-acquisition/recording device. The patented sensor cell is a three-electrode, controlled-potential device integrated on a thin disk of hydrated solid polymer electrolyte enclosed in a plastic housing. The sensor is placed over the skin surface and continuously oxidizes excreted ethanol. The oxidation current, four electrons per ethanol molecule, is used as a direct measure of the local ethanol vapor concentration over the skin surface. The sensor is highly responsive to ethanol and unresponsive to potential interferents such as oxygen and acetone. Two thermistors imbedded in the sensor produce temperature signals that can be used to determine whether the device was in continuous contact with the skin and also compensate the signal for temperature changes.

The sensor cell is attached by cable to a battery-operated data-acquisition recording device, which samples the ethanol and temperature signals at intervals and stores days to weeks of data according to a programmable protocol. The device records 8 d of data, sampling at 2-min intervals (up to 20 d, sampling at 5-min intervals). The data-acquisition/logic circuit is programmed, and data is down-loaded with an IBM-compatible computer with access through an RS232 port. A potentiometer calibrates the device.

In Vitro Measurement of Ethanol With TAS

In vitro studies with TAS show high sensitivity and specificity for ethanol. The sensor produces a current (i) related to the amount of ethanol oxidized by the electrode. The theoretical response of the sensor may be calculated from Equation [1], derived from Fick's law, which estimates the sensor current when measuring ethanol vapor at 37°C at equilibrium.

$$i = nFACD/L \qquad [1]$$

In the equation: $n = 4$ Eq/mol ethanol; D = diffusion coefficient for ethanol in air (≈ 0.1 cm^2/s); C = ethanol concentration (mol/cm^3); L = diffusion path length (0.75 cm); A = sensor cross section (3.88 cm^2); and F = Faraday's constant (96,500 Coul/Eq). The equation predicts that a sensor cell should generate a signal of about 12 μA/mg/dL ethanol at a distance of 0.5 cm from the solution. This calculated steady-state response is in general agreement with the observed values measured in vitro without a barrier (diffusion-limiting) membrane (10–12 μA/mg/dL).

A series of studies exposing the TAS sensor to potential interferents found in the body were conducted to determine the effect of such substances on the signal response. No interference with the TAS alcohol signal was found for oxygen (0–20%), carbon dioxide (0–5%), or acetone (100 mg/dL). For carbon monoxide, concentrations of 3000 ppm produced a small interfering signal corresponding to 10 mg/dL ethanol. Such high carbon monoxide concentrations would not be encountered by the sensor transdermally.

Figure 2 shows the signal of three sensors exposed to solutions of different concentrations of ethanol in water. The sensors are held in a temperature-controlled plastic holder and allowed to equilibrate. The graph of the steady current against ethanol concentration is linear over the tested concentration range of ethanol ($R = 0.99$). Experiments with higher ethanol concentrations show linearity of signal response up to approximately 300 mg/dL. TAS is calibrated by potentiometer adjustment to read 500 for a 100-mg/dL ethanol solution and for the particular diffusion-limiting barrier membrane used in the TAS. This value is the projected in vitro–transdermal ratio and allows the data to be graphed on the same axis. The response time for large, rapid changes in ethanol concentration is typically on the order of several minutes and is dependent upon the thickness of the diffusion-limiting barrier membrane.

In Vivo Assessment
of Alcohol Consumption in Humans

We tested the TAS on three groups of human subjects: subjects ingesting ethanol under laboratory conditions, intoxicated subjects with high BACs presented for detoxification in a hospital setting, and sober subjects not ingesting alcohol. The purpose of the study was to establish the magnitude of the transdermal alcohol signal with various sensor configurations, to determine sensitivity to various amounts of ingested ethanol, and to compare the response characteristics of the TAS signal with the BAC over time as measured by breathalyzer. Human studies were approved by the Human Research Committee of Roger Williams Medical Center, and informed consent was obtained from all subjects. Studies were conducted in accordance with *Recommended Council Guidelines on Ethyl Alcohol Administration in Human Experimentation.*[25]

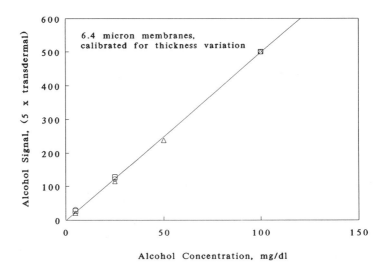

Fig. 2. Graph of the in vitro, steady-state response of TAS current to ethanol solutions of various concentrations and projected at 5× transdermal response. □ = Sensor 30; △ = Sensor 32; ○ = Sensor 35; — = Best fit; and R = 0.996.

Monitoring known alcohol consumption using the TAS indicates that the TAS alcohol signal generally follows the time course and amplitude of the BAC vs time curve. Sensor placement studies show that a robust transdermal alcohol signal is received over multiple skin areas, including the anterior abdomen, chest, and upper and lower extremities; subjects reported optimal comfort and acceptance with the TAS attached to the upper extremity. A typical TAS curve for subjects ingesting 0.75 mL of absolute ethanol per kg body weight is shown in Fig. 3, which displays the TAS signal in μA (multiplied by 40 or 80 to allow comparison with the BAC on the same axis), BAC in mg/dL, and temperature in °C, all on the Y-axis. Figure 4 shows the TAS output from a subject receiving 0.65 mL absolute ethanol per kg and 1.40 ml/kg on two consecutive days. The event marks (+) denote device placement, beginning of alcohol consumption, and device removal, respectively. TAS signals show a similar pattern and relative amplitude as the BAC. The temperature signals accurately identify times of device application and removal.

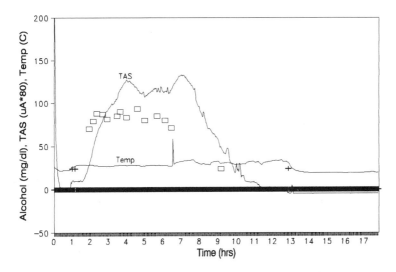

Fig. 3. TAS output (#32) of an experimental session on a human subject (#011). Subject consumed 0.75 mL absolute ethanol per kg body wt. + = Events; and □ = BAC.

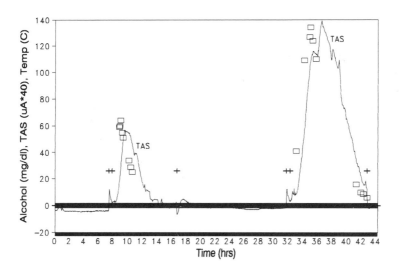

Fig. 4. TAS output (#34) from a subject (#028) consuming ethanol (0.65 ml/kg and 1.40 ml/kg) on two successive days. + = Events; and □ = BAC.

To characterize the relationship between the TAS signal and the BAC, we examined several parameters of the respective amplitude-time curves, including peak height, time to peak, area under the curve (AUC), and terminal slope in 10 nonalcoholic subjects (eight males and two females) receiving 0.75 mL of absolute ethanol per kg body wt. Pharmacokinetic parameters were calculated using the PC version of MKMODEL.[26] Means, *t*-tests, and Pearson's correlations for variables were calculated using Statistical Package for the Social Sciences (SPSS).

For the 10 subjects, BAC and TAS curve-peak amplitudes were correlated across subjects ($r = 0.61$ and $p < 0.02$) and areas under the BAC and TAS curves were highly correlated ($r = 0.90$ and $p < 0.001$). Terminal slopes of the BAC-time and AUC-time curves were not correlated and were significantly different by *t*-test ($p < 0.03$). To determine across-device, within person correlations, we examined the relationship between TAS signals and BAC for two devices worn simultaneously in nine of the subjects. Time to TAS peaks for the two devices were highly correlated ($r = 0.86$ and $p < 0.001$), as were TAS peak heights ($r = 0.71$ and $p < 0.01$) and TAS AUCs ($r = 0.94$ and $p < 0.001$). No significant correlations were obtained for TAS parameters and height, weight, age, gender, or quantity and frequency of usual alcohol consumption.

The peak values of concentration vs time curve derived from the TAS are right-shifted compared to the time vs concentration curve derived from breathalyzer data. Mean time to peak was 71 ± SEM (7 min) for the BAC curve and 107 ± SEM (12 min) for the TAS curve. The mean difference in onset times was significantly different ($p < 0.02$ by *t*-test). The threshold sensitivity for TAS in measuring ingested ethanol is approximately a BAC of 15 mg/dL, corresponding to approximately one to two standard drinks for most subjects.

TAS is also capable of tracking the pattern of BAC in intoxicated alcoholic subjects with high BACs. Figure 5 shows ethanol concentration vs time curves from an intoxicated alcoholic subject. The event marks indicate TAS application and removal, respectively. The TAS signal follows a similar pattern to the BAC curve, with a lag period of up to 120 min typically observed in the peak TAS signal and longer lags observed for the time to zero for the TAS curve. The results

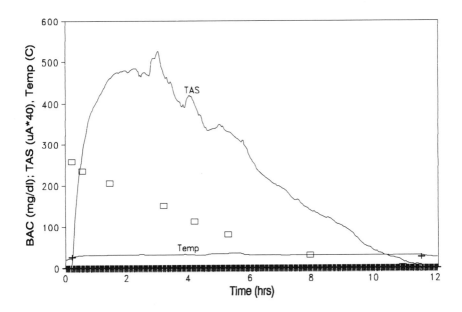

Fig. 5. TAS output (#34) from an intoxicated alcoholic subject. Device was worn for the first 12 h after inpatient hospitalization. + = Events; and □=BAC.

indicate that the TAS can track high BACs (the current detection limit is set at 800 mg/dL) and measure transcutaneous ethanol in highly intoxicated alcoholic subjects.

Initial data suggests lack of false-positive TAS signals in sober subjects wearing the TAS for periods of up to several days. We were concerned that patients with hepatic or renal failure might produce volatile blood-borne substances that could produce false-positive sensor responses, however, no false-positive TAS ethanol signals were observed in sober subjects with liver or renal disease.

Discussion

The pharmacokinetics of alcohol in humans is complex and depends upon many factors, including the amount and type of alcohol ingested, presence of food in the stomach, GI motility, hepatic blood flow, age, gender, body mass, body water, and capacities of the various alcohol dehydrogenase isoenzymes.[27] Although the volume of distribution of ethanol is generally considered to equal the body water, there are arteriolar-venous blood differences and urine ethanol concentration lags behind that of blood.[28] Several pharmacokinetic models have been used to describe the concentration of ethanol in body fluids. A one-compartment pharmacokinetic model with zero order elimination kinetics is most often applied to the analysis of BACs. The model is best applied under conditions of rapid absorption and moderate blood ethanol levels.[29,30] Multiple distribution compartments and Michaelis-Menten, first-order elimination kinetics in the model results in a better fit for low BACs and accounts for multiple alcohol dehydrogenase isoenzymes and for first-pass and other absorption effects.[30]

Measurements of human transcutaneous ethanol concentrations find approximately equal concentrations in blood and other body fluids when corrected for evaporation and water content.[13–15] However, Brown (1985) simultaneously determined ethanol in perspiration and blood and found significant differences in terminal rate constants for BAC (estimated by breathalyzer) and transcutaneous ethanol concentration.[31] He suggests that ethanol in perspiration is either not in equilibrium with blood or that skin distribution of ethanol is delayed. This discrepancy may be accounted for by the likelihood that a significant fraction of measured transdermal ethanol emanates from diffusion through the skin rather than excretion into sweat. Measurements of ethanol flux through excised samples of hydrated human stratum corneum clearly demonstrate that ethanol easily diffuses through human skin.[16]

Our studies on controlled alcohol consumption using TAS also show significant differences in terminal slopes of TAS curves and breath alcohol curves. In addition, the TAS ethanol curve lags behind the BAC curve. The peak values of concentration vs time curve derived from the TAS are rightshifted by approximately 30 min as compared to the time vs concentration curve derived from breathalyzer data. The existence of a delay in the TAS peak and a threshold sensitivity for the presence of ethanol suggests the existence of a pharmacokinetic compartment in skin that affects the time course and magnitude of transdermal ethanol excretion/diffusion; also proposed by Brown (1985).[31] Urinary alcohol vs time curves show a similar lag when compared to blood levels.[28] In spite of these differences, the TAS alcohol signal generally follows the pattern and amplitude of the BAC vs time curve.

Currently, absolute values for the BAC can be approximated but not directly derived from the transdermal ethanol signal because the transport factors necessary for such calculations are not available, e.g., the contribution from perspiration vs skin diffusion and the ethanol diffusion coefficient for the sensor barrier membrane. Yet, the TAS alcohol signal shows a high degree of correlation to the BAC when calibrated to standardized ethanol solutions in vitro, and the magnitude of the TAS response is directly related to the dose of ingested alcohol in vivo. The TAS ethanol signal is influenced by the geometry of the sensor in relation to the skin, the rates of excretion and/or diffusion through the skin, and evaporation. A breathalyzer sampling breath alcohol vapor similarly requires calibration according to the blood–breath partition ratio.[32]

Wearing the TAS was well tolerated by all sober and intoxicated subjects. No skin irritation or rashes developed in skin over the area of the sensor. The site preferred by subjects was the inner surface of the lower arm or outer surface of the upper arm. Devices were worn in a single location for periods of up to 7 d without discomfort.

In order to accurately monitor alcohol consumption over time, the method of monitoring must be tamper-proof, and the compliance rate of subjects must be high. Individuals whose drug or alcohol use is monitored for clinical or forensic purposes may go to great lengths to falsify or invalidate the results.[33] The TAS thermistor temperature signal may be used as an indicator of compliance, however, the temperature may be influenced by a number of factors, including ambient temperature. Additional measures are under consideration to ensure that the device is properly placed and tamper-resistant and to detect tampering. These include a galvanic skin-response detector incorporated into the sensor and the use of adhesives that secure attachment of the TAS to skin.

Conclusions

The Giner, Inc. TAS shows significant promise for determining the time course and relative amount of alcohol consumption on a continuous, real time basis (Table 1). The device accurately tracks transdermal ethanol, which is related to the BAC. The device is currently being utilized in clinical studies on larger sample sizes and

Table 1
Giner TAS Assessment of Alcohol Consumption

Accuracy: Accurately determines pattern of use but is semiquantitative as to amount. Measurement of absolute values of BAC should improve with better calibration.

Ease of Use and Cost: Device will be easy to apply and use by essentially all personnel after minimal training. Data analysis and interpretation require a personal computer with a serial port. Cost of the device is projected at $800 per unit, with each unit utilized for testing many subjects.

Utility for Various Settings: Especially valuable as research tool to assess consumption in cooperative subject . Utility in clinical and forensic settings will depend on improved tamper-resistance.

Latency and Retrospective Accuracy Capabilities: TAS responds within minutes of application. It is useful for prospective determinations only.

Acceptability: Initial studies indicate high degree acceptability by both practitioners and clients. Field trials are ongoing to assess acceptability across different subjects and settings.

Unique Benefits: Only method of providing continuous measure of alcohol consumption for extended time periods.

Problems: Has threshold limit of detection of approximately 15–20 mg/dL for BAC.

Water-Resistance: Needs to be improved to allow continuous wearing of device while swimming or bathing. Tamper-resistance also needs to be improved. Further miniaturization of the device is in progress.

---

across subjects with different skin characteristics. Additional testing must be performed to better elucidate the clinical pharmacology of transcutaneous ethanol and its relation to BAC for better quantitation of alcohol consumption. In addition, further studies are needed to test reliability, specificity, and acceptance of the method in different groups of individuals and over a range of research and clinical applications. Successful development of such a detector has important implications for alcohol research and treatment, including validating other methods of assessing alcohol consumption and monitoring treatment outcome, treatment compliance, and individuals who are mandated to remain abstinent from alcohol.

References

[1]Chapter 11, this volume.

[2]L. C. Sobell, S. A. Maisto, M. B. Sobell, and A. M. Cooper (1979) Reliability of alcohol abusers' self-reports of drinking behavior. *Behav. Res. Ther.* **17,** 157–160.

[3]B. S. McCrady, T. J. Paolino, and R. Longabaugh (1978) Correspondence between reports of problem drinkers and spouses on drinking behavior and impairment. *J. Stud. Alcohol* **39,** 1252–1257.

[4]T. F. Babor, R. S. Stephens, G. A. Marlatt (1987) Verbal report methods in clinical research on alcoholism: Response bias and its minimization. *J.Stud. Alcohol* **48(5),** 410–424.

[5]T. J. O'Farrell and S. A. Maisto (1987) The utility of selfreport and biological measures of alcohol consumption in alcoholism treatment outcome studies. *Adv. Behav. Res. Ther.* **9,** 91–125.

[6]L. C. Sobell, M. B. Sobell, G. Leo, and A. Cancilla (1988) Reliability of a timeline method: Assessing normal drinkers' reports of recent drinking and comparative evaluation across several populations. *Brit. J. Addict.* **83,** 393–402.

[7]R. K. Fuller, K. K. Lee, and E. Gordis (1988) Validity of self-report in alcoholism research: Results of a veterans administration cooperative study. *Alcoholism: Clin. Exp. Res.* **12,** 201–205.

[8]H. Stibler and S. Borg (1981) Evidence for reduced sialic acid content in serum transferrin in male alcoholics. *Alcoholism: Clin. Exp. Res.* **5,** 545–549.

[9]R. S. Ryback, M. J. Eckhardt, B. Felsher, and R. R. Rawlings (1982) Biochemical and hematologic correlates of alcoholism and liver disease. *JAMA* **248,** 2261–2265.

[10]P. Cushman, G. Jacobson, J. J. Barboriak, and A. J. Anderson (1984) Biochemical markers for alcoholism: sensitivity problems. *Alcoholism: Clin. Exp. Res.* **8,** 253–257.

[11]S. Takase, A. Takada, M. Tsutsumi, and Y. Matsuda (1985) Biochemical markers of chronic alcohol. *Alcohol* **2,** 405–410.

[12]E. Nyman and A. Palmlov (1936) The elimination of ethyl alcohol in sweat. *Skandinavisches Archiv fur Physiologie* **74,** 155-159.

[13]G. L. S. Pawan and K. Grice (1968) Distribution of alcohol in urine and sweat after drinking. *Lancet* **2,** 1016.

[14]S. W. Brusilow and E. H. Gordis (1966) The permeability of the sweat gland to non-electrolytes. *Am. J. Disease Children* **112,** 328–333.

[15]D. J. Brown (1985) Pharmacokinetics of alcohol excretion in human perspiration. *Methods Findings Exp. Clin. Pharmacol.* **7(10),** 539–544.

[16]R. J. Scheuplein and I. H. Blank (1971) Permeability of the skin. *Physiol. Rev.* **51(4),** 702–747.

[17]M. Phillips (1980) An improved adhesive patch for long-term collection of sweat. *Biomater. Med. Devices Art. Organs* **8(1)**, 13 –21.

[18]M. Phillips and M. H. McAloon (1980) A sweat-patch test for alcohol consumption: Evaluation in continuous and episodic drinkers. *Alcoholism: Clin. Exp. Res.* **4 (4)**, 391–395.

[19]M. Phillips (1982) Sweat patch test for alcohol consumption: Rapid assay with an electrochemical detector. *Alcoholism: Clin. Exp. Res.* **6(4)**, 532–534.

[20]E. L. R. Phillips, R. E. Little, R. S. Hillman, R. F. Labbe, and C. Campbell (1984) A field test of the sweat patch. *Alcoholism: Clin. Exp. Res.* **8(2)**, 233–237.

[21]A. H. Parmentier, M. R. Liepman, and T. Nirenberg (1991) Reasons for failure of the alcohol sweat patch. *Alcoholism: Clin. Exp. Res.* (abstract), in press.

[22]A. W. Jones (1978) A rapid method for blood alcohol determination by headspace analysis using an electrochemical detrector. *J. Forensic Sci.* **23**, 283–291.

[23]K. A. Gibb, A. S. Yee, C. C. Johnston, S. D. Martin, and R. M. Nowak (1984) Accuracy and usefulness of a breath alcohol analyzer. *Ann. Emerg. Med.* **13(7)**, 516–520.

[24]K. Dubowski (1976) Recent developments in breath alcohol analysis. *Proceedings of the 6th Int. Conference on Alcohol, Drugs and Traffic Safety.* S. Israelstam and S. Lambert, eds. (ARF of Toronto, Toronto, Ontario).

[25]NIAAA (1989) *Recommended Council Guidelines on Ethyl Alcohol Administration in Human Experimentation.* National Advisory Council on Alcohol Abuse and Alcoholism.

[26]N. Holford (1990) MKMODEL, Version 4. Biosoft, Cambridge, UK.

[27]H. Kalant (1971) Absorption, diffusion, distribution and elimination of alcohol, *The Biology of Alcoholism*, B. Kissin, H. Begleiter, eds. (Plenum, New York).

[28]R. D. Batt (1989) Adsorption, distribution and elimination of alcohol, in *Human Metabolism of Alcohol*, vol.1, K. E. Crow, R. D. Batt, eds. (CRC, Boca Raton, FL).

[29]E. M. P. Widmark (1932) *Die theoretischen Grundlagen und die praktische Verwendbarkeit der gerichtlichmedizinischen Alkoholbestimmung* (Urban and Schwarzenberg, Berlin).

[30]J.-P. von Wartberg (1989) Pharmacokinetics of alcohol, in *Human Metabolism of Alcohol*, vol.1, K. E. Crow and R. D. Batt, eds. CRC Press, Boca Raton, FL.

[31]D. J. Brown (1985) A method for determining the excretion of volatile substances through skin. *Methods Findings Exp. Clin. Pharmacol.* **7(5)**, 269–274.

[32]K. Dubowski (1976) Human pharmacokinetics of ethanol 1. Peak blood alcohol concentrations and elimination in male and female subjects. *Alcohol Tech. Rep.* **5**, 55–63.

[33]R. M. Swift, P. Camara, and W. Griffiths (1991) Use of toxicological analysis in the identification of drugs of abuse. *Medical Psychiatric Practice*, vol.1, A. Stoudemire and B. Fogel, eds. (APA Press, Washington, DC).

SUMMARY

Measures
of Alcohol Consumption
in Perspective

John P. Allen, Raye Z. Litten,
and Raymond Anton

Introduction

This chapter offers a general summary and perspective on measures of alcohol consumption. In it, we address five issues:

1. Differences between verbal and biochemical techniques to assess alcohol consumption,
2. Features that are desirable in measures of consumption,
3. Comparisons of specific techniques according to these standards,
4. Possible strategies for measuring consumption in various settings, and
5. Needs for further research.

The individual chapters of this text have presented an array of verbal and biochemical measures. In selecting the techniques to include in the text, we limited ourselves to those that allow assessment of consumption some time after drinking has actually occurred. Clearly, techniques such as analysis of breath, blood, and urine serve as straightforward measures of the presence of alcohol in the body. They do not, however, indicate the maximum amount of alcohol that may have been consumed, and most importantly, they provide information for only a very short period after drinking has occurred.

From: *Measuring Alcohol Consumption*
Eds.: R. Litten and J. Allen ©1992 The Humana Press Inc.

Description of Specific Techniques

The text has reviewed and briefly discussed the literature on the following nine measures of consumption:

1. Traditional self-report measures
2. The timeline follow-back (TLFB) procedure
3. Computer-assisted self-reports
4. Collateral (C) reports
5. γ-Glutamyl transferase (GGT)
6. Carbohydrate-deficient transferrin (CDT)
7. 5-Hydroxytryptophol (5-HTOL)
8. Acetaldehyde adducts (AA) and
9. Transdermal devices.

Traditional Self-Report Measures

These are the oldest and most flexible techniques for measuring past alcohol consumption. Self-reports of drinking generally take the form of either a written questionnaire or a structured interview asking the patient to offer a summary description of his or her drinking behavior in terms of quantity of ethanol consumed during each typical drinking episode over some time frame in the past. Information on specific brands of alcoholic beverage is rarely solicited, but the patient is often asked to give responses in terms of standard drink equivalents, e.g., 12 oz beer = 5 oz ordinary wine = 3 oz of fortified wine = 1 standard cocktail or a shot and a half of hard liquor.

Timeline Follow Back

In this procedure, the patient is generally given a blank calendar covering some period of time in the past. The interviewer and patient then annotate particular days as having personal or general significance, such as holidays, days in jail, birthdays, and so forth. These events serve as memory cues in helping the patient recall the specifics of drinking. The patient is next asked to identify his or her most common drinking pattern, including amount and type of alcoholic beverage by time periods of the day (usually defined as meal-to-meal, before breakfast, and after dinner), and to mark the calendar indicating the time frames for this most typical drinking pattern. Following this, the patient is asked about other types of recurrent, consistent drinking patterns, such as weekend or holiday drinking styles. Hopefully, the calendar can be largely annotated by filling it in with fairly standard drinking patterns. Finally, the patient is urged to estimate the drinking that occurred on the days of the calendar that have not been filled in.

The TLFB approach differs from traditional self-report measures by seeking a high degree of specificity in the recollection rather than asking the patient to average his or her drinking over some time frame. TLFB also yields considerably more information than traditional summary measures of drinking. These include peak blood alcohol levels, periods of heavy consumption, periods of abstinence, and drinking patterns. Computer programs exist to score these variables and to collect data from patients.

Computer-Assisted Self-Reports

As the name implies, this method involves having the patient report on drinking through a computer program rather than giving this information directly to an interviewer or completing a written form. Computer-assisted methods typically decrease interviewer time in ascertaining the information. Equally important, the computer program may include branching algorithms that systematically probe certain kinds of responses for which more information is needed. The computer may also score variables such as response latency, which has been found to be associated with patient veracity in responding.

Collateral Reports

This technique has also been used to assess drinking by the patient. Typically, this information is used as a means of corroborating the patient's own report of drinking. The questions employed in C reports are often more general than those included in direct patient reports since collaterals are rarely able to report accurately on specific amounts of drinking and also rarely have consistent contact that would allow them to observe all drinking behavior of the patient.

γ-Glutamyl Transferase

GGT appears to be the most commonly employed laboratory test indicator of possible chronic heavy alcohol use. GGT is an enzyme produced in liver cells; and secretion of GGT is often elevated if liver cells are exposed to a significant amount of alcohol over an extended period of time. The amount of GGT can be measured in the peripheral venous blood and readily analyzed in most clinical laboratories. The results can be reported in a clinically useful time frame. Although the cost of GGT measurement may be slightly higher than other liver function tests in some laboratories, it is not excessively expensive. In some medical centers, it is available in a battery of other tests done with automated procedures and therefore can be easily obtained during routine physical screening. The most significant problem with the GGT, despite its moderate sensitivity (40–60%), is its low specificity. This is a particular problem in relationship to liver pathology of other etiologies and its induction by certain pharmaceutical compounds (*see* chapter 5).

Carbohydrate-Deficient Transferrin

CDT is one of the newest and perhaps most promising of the biological markers of alcohol consumption. To date, a significant amount of data has been accumulated and published on its validity and reliability. Transferrin, a protein with attached carbohydrate groups, is produced under normal conditions by the liver cell and released into the blood. The protein binds iron in the blood and transports it to other tissues, where it transfers it to other proteins, such as enzymes, hence the name *transferrin*. Under conditions of moderately heavy alcohol use, the liver cell releases transferrin without the normal amount of carbohydrate groups attached, and therefore, in the peripheral blood, an abnormal type of transferrin, CDT, can be measured.

Currently, the main technological difficulty with the procedure is the separation of CDT from naturally occurring transferrin, which continues to be present even under conditions of heavy alcohol use. This procedure is beyond the capability of most clinical laboratories because of the high level of requisite technical expertise and expensive instrumentation. Undoubtedly, continued research on CDT will lead to more practicable and cost-effective measurement techniques.

It is likely that further development will render this more readily available at medical centers and reference laboratories around the world. Within a number of years, if the data obtained to date continue to stand the tests of replication and generalizability, the measurement of CDT may become available and routine in many clinical settings.

5-Hydroxytryptophol

This method is even newer than CDT and as such, has a smaller data base from which to draw conclusions about its general utility as a test of alcohol consumption. A metabolite of serotonin, 5-HTOL is excreted in the urine after conjugation with glucuronic acid. After the acute ingestion of alcohol, 5-HTOL levels increase (*see* chapter 7). This compound can be extracted from the urine and measured with relatively sophisticated instrumentation. It has been observed that if the amount of 5-HTOL is compared to the amount of another naturally occurring metabolite of serotonin, a higher sensitivity for the test can be achieved.

From currently available data, it appears that 5-HTOL is most useful in ascertaining acute ingestion of alcohol and therefore may be limited in its ability to screen for heavy continued use in a retrospective fashion. It may best be used as either a prospective marker of relapse or in screening for recent consumption. In any case, these issues will come into greater focus as more data evolve regarding its sensitivity and specificity. Additionally, given the nature of its measurement, its transportability out of the research laboratory and into the clinical setting will need further development.

Acetaldehyde Adducts

Several studies have shown that acetaldehyde, a highly reactive metabolite of ethanol, binds to a variety of proteins, including plasma proteins and hemoglobin, to form a stable protein–AA. Since most of the acetaldehyde comes from ethanol metabolism, the amount of protein–AA formed reflects the quantity of alcohol consumed over a period of time. Recent results have reported increased serum levels of protein–AA in alcoholics. Current research is directed toward improving the methodology, such as the appropriate conditions for sample storage and the development of more specific antibodies to protein–AA. Also, researchers are evaluating the potential problem of endogenous acetaldehyde formation, particularly in nonalcohol-related liver disease, which leads to elevated levels of protein–AA in the absence of drinking.

Transdermal Devices

Significant progress has been made to develop a clinically operative transdermal device. Two models have been described in this volume, the transdermal dosimeter TDD, or "sweat patch," and the transdermal alcohol sensor/recorder (TAS). The TDD determines amount of alcohol in the transepidermal fluid. TAS is at an earlier stage of development and measures the ethanol vapor at the surface of the skin. A unique feature of the latter device is its ability to record the time of the alcohol vapor measurement. Both devices are noninvasive. These promising devices are currently undergoing further design modifications and field testing.

Interpretation of Tests
to Measure Consumption

In order to have a common language and method of comparing the utility of tests in the diagnosis or categorization of alcohol use, abuse, or dependency, a conventional system has been utilized by researchers. The terms of this system need to be understood and applied to the evaluation of past and future methods developed to measure alcohol consumption. These types of analyses have been applied to other diagnostic tests.[1–3]

In general, both researchers and clinicians would like to know how confidently they can determine a certain level of alcohol use from a given test level and how certain they can be that alcohol use was the only factor that could have caused a measurement to reach a certain (abnormal) level. The former need— restated as the ability of a test to identify people who have the clinical characteristic of concern, in this case, a given level of alcohol use—is known as sensitivity. The second need, restated as the ability of a test to identify only those affected by the given condition, is known as specificity. For example, the ideal test for a given level of alcohol use, say, five or more drinks each day, would be 100% sensitive if it was "abnormal" for everyone who drank at least at that level. If it was abnormal in only one half of the patients who drank at that level, then the sensitivity of that test would be 50%. In addition, if the test was abnormal only in people that drank at that level and in no other condition (such as liver disease), then the test would be 100% specific. Conversely, if one half of nondrinking people with another condition (such as liver disease) or with no identifiable condition at all (normal controls) had an abnormal test, then the test would have a specificity of 50%.

When a test does not correctly identify a person who has the given condition, e.g., alcohol abuse, the result would be considered a false-negative. When a test incorrectly identifies a person without the given condition, e.g., alcohol abuse, as having the condition, the result would be considered a false-positive. A test that has high sensitivity but low specificity will identify many more false-positives when the prevalence of the condition being tested for in a given population is low. For instance, if a test is performed on 100 people and the condition (alcohol abuse) is expected to be present in 90 of them (as may occur in a substance-abuse treatment center) and if the test has 100% sensitivity (no false-negatives) but has only 90% specificity (one out of 10 false-positives), then the results of the testing would show 91 people positive for the condition (alcohol abuse), of whom 90 will have the condition and one will not. On the other hand, if the same test is performed on 100 people and the condition (alcohol abuse) is expected to be present in only 10 of them (as may occur in a medical outpatient setting), then the results of the testing would show 19 people positive for the condition; 10 will have the condition, and nine will not. As can be seen, in the first condition, the specificity (90%) of the test does not detract much from its clinical relevance. In the second condition, where the base rate (prevalence) of the condition is lower, the lack of 100% specificity of the test could have significant clinical implications, e.g., misidentifying a relatively large number of people.

Given the above, it is clear that the acceptability (validity and clinical utility) of a given test procedure may vary depending on the population in which it is being utilized and the purposes of the test. Clearly, in a screening situation where the base rate of alcohol abuse is low (such as in airline pilots), tests that are sensitive but also highly specific may be required. In contrast, in a treatment followup setting where alcohol abuse is the condition being treated and other interfering conditions (such as liver disease) have been previously ruled out, a test with a lower specificity may be acceptable.

Another issue regarding tests of all kinds deals with the concepts of validity and reliability. The issue of validity has to do with the ability of a test to actually measure what it is intended to measure. Another term for this is *accuracy*. The best and perhaps most efficient method of determining the accuracy or validity of a certain procedure is to compare the results of the new procedure against an established and presumably accurate (valid) procedure. Since there are no "gold standards" against which to measure a new test of alcohol consumption, this can only be approached by using the best tests that are available or combining several procedures and requiring agreement on several other measures as an indication of validity.

In addition to accuracy, a test must be precise or reliable. Precision is the ability of a given test to produce the same result when a sample (subject) is measured on several occasions. In a sense, precision is a measure of the "noise" in the testing procedure. In a biological assay situation, precision can be affected by the technician, the quality of the reagents, the assay conditions (such as temperature, pH, time, and so forth), and the measurement instrument. In a clinical situation, precision can be affected by the state of the patient (level of intoxication or day of withdrawal), the skill and training of the interviewer, the setting (screening vs treatment), the ease of use of the instrument, and the time span over which the information is gathered (recent vs remote past).

Obviously, a test is much more acceptable if it gives accurate and precise information. One characteristic of the ideal test for alcohol consumption is the generalizability of the procedure. When a test is utilized in a new setting or population, it should be tested for both accuracy and precision. A test that has less "noise" inherent to the procedure should be more accurate and therefore more easily generalizable. Commonly, by their nature, biological laboratory tests are more easily controlled, and have less "noise" and hence greater accuracy characteristics than nonbiological tests.

Finally, the amount of alcohol consumption that one wishes to detect will influence the sensitivity characteristics of the test. In general, a test that will detect the condition of any alcohol use will need to be very sensitive. However, in order to get the sensitivity that is desirable, most often specificity is sacrificed. As a rule, as the sensitivity of a test increases, specificity decreases. A mathematical method has been developed that takes this relationship into account in aiding the researcher in selecting the optimum combination of sensitivity and specificity for any given test. This is referred to as Receiver Operating Characteristic, or ROC, analysis.[4] Many researchers feel that this methodology should be utilized in setting the cutoff points in test measurements for optimum detection of the condition for which the tests are being conducted.

General Comparison of Verbal and Biochemical Techniques

An overriding concern among theorists dealing with measuring alcohol consumption is the degree of accuracy that should be ascribed to traditional self-reports of drinking. Problems with such measures have been most recently discussed by Rosman and Lieber.[5] The extent to which these problems might also diminish other verbal techniques, such as the TLFB method, C reports, and more sophisticated ways of framing questions, is unclear. Since the validity of verbal measures of alcohol consumption may be reduced by memory problems and unwillingness to acknowledge drinking, physiological measures of consumption may, in some instances, be preferred.

It has not been our intention in preparing this text to in any way compete verbal techniques and the biochemical techniques with each other. Nevertheless, it is interesting to compare and contrast them on six broad dimensions of concern:

1. Sources of Error
2. Credibility
3. Flexibility
4. Scaling of Results
5. Expenditure of Effort, and
6. Applicability to Varying Settings.

Sources of Error

Although the sources of contamination differ across all nine techniques, certain ones are primarily associated with verbal measures and others with biochemical measures. The particular threats to validity of the verbal measures tend to be problems with memory, difficulties in understanding questions, performing mental calculations to quantify drinking, and intentional dissimulation. Chapter 1 addresses these and other potential contaminants and offers some promising remedies. Sources of error for biochemical measures more clearly conform to traditional concerns with sensitivity and specificity. Often, a rather large amount of alcohol must be consumed over a relatively long period of time, temporally proximal to the specimen collection, for biochemical measures to be elevated. On the other hand, nonalcohol-related liver pathology or other disease processes may result in a false-positive test.

Credibility

Ascertainment of alcohol consumption is of importance to physicians, alcoholism treatment program evaluators, law-enforcement officials, members of the patient's family, the patient, and so forth. Many of these individuals lack a research background or in any event, are unaware of research results on the validity on the techniques. Biochemical techniques and to a lesser extent, C reports are likely to provide results that are more convincing to such groups, regardless of the accuracy of the technique. This is not a trifling issue. For example, physicians and perhaps patients may well be more concerned with drinking problems if laboratory tests suggest problematic consumption than if the evidence of heavy consumption is based primarily on a verbal report. Outside program or treatment study reviewers also will probably be more impressed with decreases in biochemical measures of drinking posttreatment than with decreases of self-acknowledged drinking.

Flexibility

Verbal techniques for measuring consumption are generally more flexible than biochemical markers, since they can capture a range of different time periods and inquire about varying patterns, levels, and consequences of consumption biochemical measures, but tend to be highly standardized. It is conceded, however, that computerized measures of consumption and formal pencil-and-paper measures are also rather standardized. Nevertheless, the setting for the evaluation, the patient's psychological set, rapport with the individual giving instructions, and so forth keep these techniques from achieving the same level of standardization achievable by biochemical measures.

Nature of Results

The greater flexibility of verbal measures typically allows them to yield data on level of consumption in the form of a continuous variable. Biochemical measures as currently utilized, however, generally yield only dichotomous results, i.e., drinking vs abstinence or drinking at a moderate level vs drinking excessively. For certain purposes, a dichotomous-dependent variable is probably adequate. These might include determining whether a patient is drinking at such a level that she or he risks development of alcohol-related medical problems or whether a patient following alcoholism treatment is now totally abstinent. For most purposes, however, greater precision in the measures of drinking is usually desirable. Statistical power is higher with continuous variables and hence, fewer cases are required in research for which the dependent variable is alcohol consumption. So too, continuous variable measurement allows more accurate contrast of outcomes across competing alcoholism interventions.

Expenditure of Effort

In comparing the various techniques to assess alcohol consumption, it also is possible to distinguish the stage of the evaluation that requires the greatest expenditure of effort and resources. In general, verbal measures are more difficult to administer but simpler to score and interpret. Biochemical techniques, on the other hand, are relatively easy to administer (collecting the specimen), but require more effort, technical skill, and instrumentation to perform the test. Use of mathematically sophisticated means of combining results from separate measures of consumption moves the major effort from administration and scoring to interpreting results.

Applicability to Various Settings

As noted in the foreword of this text, there are a variety of contexts and reasons for assessing alcohol consumption. These range from a general community setting, such as performing health-risk appraisals in a shopping center, to performing a methodologically complex efficacy study on alcoholism treatment. Verbal procedures, especially traditional self-report techniques, are simple, rapid, and require little technical expertise. In contrast, biochemical techniques commonly require very sophisticated laboratory tests that are usually not available in a general community setting or conversely, require that the results be sent elsewhere for evaluation. This results in delaying feedback and advice to be given to patients.

Comparisons Among Specific Techniques

Tables 1 and 2 compare various techniques for measuring alcohol consumption. The techniques rated in the tables are those discussed at length in this volume. Certain other biochemical measures have been alluded to in the text but are not included here. Mean corpuscular volume (MCV) is not included because of its low sensitivity and the delay in its response to abstinence. The two liver transaminases, aspartate aminotransferase (AST) and alanine aminotransferase (ALT), as well as high-density lipoprotein, also are omitted because of their poor sensitivity and/or specificity. Red blood cell (RBC) acetaldehyde and cysteine show promise as markers of consumption but have yet to receive research sufficient to allow conclusions to be formulated about them. Finally, β-hexosaminidase has also been excluded because of the need for further evaluation. The six criteria on which the techniques are compared deal with features that are desirable in a marker of alcohol consumption and are discussed below.

Window of Assessment

This term refers to the time frame following drinking that can be measured. We have distinguished the dimensions of window of assessment and accuracy as desirable properties of markers. Window of assessment deals with the latency to the measure, i.e., the time frame for postdrinking or postproblematic drinking that can be assessed. As noted earlier, body fluid measures of alcohol have such a short window of assessment that they are not considered here. Accuracy deals with the responsiveness of the measure to drinking. Verbal reports offer advantages because their window of assessment extends

Table 1
Verbal Measures of Alcohol Consumption

Category	TLFB	Self-reports	C reports	Computerized reports
Window of assessment	Mo/Yr	Yr	Undetermined	Yr
Accuracy				
Any alcohol	High	Moderate	Moderate	Moderate
High intake	High	Moderate	Low	Moderate
Ease of use	Moderate	High	Moderate	High
Acceptability	Moderate	High	High	High
Cost	Moderate	Low	Low	Moderate
Transportability	Moderate	High	High	Moderate

Table 2
Biochemical Measures of Alcohol Consumption

Category	GGT	CDT	AA	5-HTOL	TDD
Window of assessment	Wk	Day/Wk	Undetermined	Hr/Day	Day/Wk
Accuracy					
Any alcohol	Low	Low	Low	Undetermined	High
High Intake	Moderate	Moderate/High	Moderate	High	High
Ease of Use	High	Low	Low	Low	Moderate
Acceptability	Moderate	Moderate	Moderate	Moderate	Low
Cost	Low	High	High	High	Undetermined
Transportability	High	Moderate	Low	Low	High

to periods temporally remote from the time of assessment. Certain liver function tests also reflect long-term heavy consumption, but it is often difficult to assign the specific historical period for this.

It also should be noted that most markers are retrospective in nature; that is, they estimate drinking previous to the period in which the individual is evaluated. The TDD, however, can only be used prospectively, since the patient would not likely have been given the device until he or she had already been seen the first time. The same would be true of a log or diary that the patient would keep of his or her drinking. So, too, some researchers have argued for assessing differences on biochemical tests between two points in time rather than absolute values on the tests as a means of gauging consumption.[6] It should also be borne in mind that prospective measures of drinking also may be quite reactive. They may not only assess drinking but also serve as an impetus to suppress it, since they may cause the patient to more attentively monitor drinking. Obviously, this may be an advantage clinically, but from a research perspective, it is problematic because the effect of intervention is confounded by how it is assessed.

Accuracy

Accuracy refers to a summary evaluation reflecting the correlation between values resulting from the technique and the amount consumed. It also refers to the precision of the measurement. We have subdivided this criterion further. Some techniques simply attempt to distinguish drinking from not drinking or more commonly, drinking in a problematic manner from drinking moderately or not at all. Other techniques are better able to identify high levels of drinking on a continuum. In rating the techniques against this criterion, we have tried to evaluate them against the standard the proponents of the technique seem to have sought for it. Hence, CDT is given a favorable score on the "high intake" aspect of accuracy. One to two wks of moderate to heavy consumption (five to six drinks per day) appear necessary to elevate it, and the half-life of the elevated value is approximately 2 wk. Traditional quantity frequency (QF) self-reports are given only moderate scores on both aspects of accuracy, although it appears that proper wording of questions and employment of contextual cues in interviewing may improve the accuracy of such measures.

Ease of Use

This category deals with two phenomena: difficulty involved in collecting the specimen or response and the level of difficulty involved in evaluating it, whether by laboratory techniques or interpretation of verbal response. Our summary judgement of this criterion also involves generalized assumptions about applied assessment settings. As noted above, traditional self-report measures can be readily introduced into almost any conceivable setting. Sophisticated laboratory tests and automated self-report measures are more difficult to incorporate into general public health and workplace settings.

Acceptability

Acceptability refers to the presumed willingness of practitioners to employ the measure and the likely willingness of subjects to submit to it. Markers such as blood tests involve some inconvenience and discomfort for patients. The labor intensity of the TLFB method may well make it less inviting to busy practitioners.

Cost

Direct financial burdens involved in administering and evaluating the tests as well as an estimate of practitioner labor are considered by cost. Scoring techniques are currently rather costly for techniques such as automated interviewing because additional software must yet be developed, and in many sites, personal computers would need to be purchased. Some of the newer biochemical measures, such as CDT, 5-HTOL, and protein–AA, require additional research and development expenditures before they can earn a role as routine laboratory procedures. Once they are used clinically, labor and instrument costs will probably make them relatively expensive routine laboratory procedures. However, with further development, these costs should diminish.

Transportability

Transportability is the degree of ease in transferring the technique from a research setting to an applied setting. Techniques such as the TLFB method require intensive staff training. Even with the development of lab kits to measure CDT and protein–AA, considerable precision on the part of the technician will probably still be required.

Tables 1 and 2 reflect our own summary judgements based on the data as we interpret them. Other investigators, especially advocates for the various techniques, may well disagree somewhat. Also, our summary judgments consider only the current research. GGT and traditional self-reports have been investigated extensively. Newer techniques, such as 5-HTOL and automated self-report measures, have been less subjected to research. Our views of these newer techniques could well change as more information becomes available.

Utility of Tests in Various Settings

Given the multiplicity of measures discussed in this text, with their varied methodologies and differing levels of sensitivity and specificity, the authors felt that it would be helpful to condense this information and propose strategies for use of the tests in a range of settings. The authors considered the type of information that a clinician, researcher, public safety official, or employer might desire and also the ease of use and acceptability of the procedure for that given setting.

This task is idealized in the sense that the methods could be best applied in the specific settings if the time, technology, and motivation were present. Obviously, the information provided here should be utilized with other data to choose the most appropriate procedures for a particular setting. An attempt was made to be somewhat futuristic and to incorporate into Table 3 certain tests, particularly biological ones, that are close to full clinical development and utility. As such, this review might also serve as a guide toward further examination of the utility of specific indicators in various settings.

In general, it is suggested that verbal report measures are primarily useful in clinical situations, such as treatment and evaluation settings, where the patient has little reason to deny his or her use of alcohol. Brief self-report measures are especially flexible and can be incorporated into most settings. Techniques to enhance their validity as suggested in the chapter on self-reports are recommended. Nonconsumption-based verbal measures of alcoholism such as CAGE,[7] Short Michigan Alcoholism Screening Test,[8] Structured Clinical Interview for DSM-III-R,[9] and Addiction Severity Index[10] are also often helpful. On the other hand, in instances where denial or minimizing alcohol use is

Table 3
Alcohol Consumption Measures for Specific Populations and Settings

Measurement setting	Verbal methods	Biochemical methods
Primary care		
Health screening units for		
Heavy consumption	C Reports	Liver tests/CDT
Alcoholism	C Reports	Liver tests/CDT
Medical treatment units	C Reports/TLFB	CDT/AA
Specialized medical settings		
GI clinics	C Reports/TLFB	CDT/MCV
Detox units	C Reports/TLFB	Liver tests/CDT
Trauma units	C Reports/TLFB	CDT/AA
Neurology units	C Reports/TLFB	CDT/AA/MCV
Public safety		
Transportation		5-HTOL/CDT
Medical personnel		Liver tests/CDT/AA
Probation after alcohol offenses	C Reports	5-HTOL/CDT/TDD
Work place		
Investigation of source of performance problem	C Reports	Liver tests/CDT
Adherence to alcoholism treatment	C Reports	5-HTOL/CDT/TDD
Alcoholism treatment		
Clinical effectiveness	C Reports/TLFB	5-HTOL/CDT/TDD
Clinical research studies		
Detection	C Reports/TLFB	Liver tests/CDT/AA
Treatment effectiveness	C Reports/TLFB	5-HTOL/CDT/TDD

common, such as in the workplace, medical evaluation, and substance-abuse treatment settings, a nonpatient-generated verbal report (C report) or a biologic laboratory test also seems indicated. Additionally, in the clinical treatment followup situation, where denial is often present to a higher degree, biological laboratory measurements are indicated. Standard measures of current degree of blood alcohol level, such as breath, blood, and urine tests, also play an important role in most settings, especially those in which immediate decisions regarding degree of intoxication must be made.

The other dimension that needs to be considered in choosing a test of alcohol consumption is the period of evaluation. For instance, a much different procedure must be utilized to determine lifetime alcohol consumption than one that would be utilized to screen for recent heavy use. Even within the available biological laboratory markers, there may be considerable differences in the ability of these tests to detect different lengths and amounts of alcohol use. For instance, it appears that for CDT to be detected in abnormal amounts in the blood, it is necessary for the person to have consumed about five to six drinks a day for 1–2 wk. In contrast, to detect abnormal levels of 5-HTOL in the urine, 1 day of drinking at this level is needed.

In any case, it appears that research findings will continue to expand the number of methods and tests available to determine alcohol consumption. The increasing variety will allow researchers and clinicians greater flexibility in choosing the appropriate procedure to maximize the information on which to base intervention, clinical treatment, and study outcome decisions.

Needs for Future Research

Several of the individual chapters of this text have discussed needs for future research on specific measures. Beyond this, however, there are several types of research that cut across the differing techniques. It would, for example, be interesting to perform error analyses on the various techniques. While several projects have contrasted two or more individual methods in terms of relative accuracy, little work has been done to contrast techniques in terms of the specific types of errors to which each is vulnerable. If techniques were compared in terms of the nature of the kinds of errors they tend to make, it should be possible to generate strategies to develop robust sequences of complimentary measures. For example, CDT seems to identify individuals who have engaged in a number of days of continuous moderate to heavy drinking (five to six drinks a day) during the previous couple of weeks. Patients who score in the nonclinical range of CDT values might, however, be identified through TLFB.

A large-scale study would allow such analyses using the various techniques with a highly heterogeneous sample of cases differing in drinking patterns, levels of consumption, duration of heavy consumption, liver pathology, neuropsychological status, and motives for revealing or suppressing their drinking. This type of project also would determine the areas of convergence and discrepancy among measures; and it would be helpful in establishing greater consistency in research methodology on the various kinds of markers. If one is to objectively compare alternative methods for measuring alcohol consumption, then consistent descriptors of patients, base rates of drinking in the population studied, rates of refusal in taking the test, and consistent reporting of sensitivity and specificity are needed.

Studies on verbal report measures should include exploration of how patient motivational and cognitive variables affect validity. Such research should also evaluate the extent to which the validity of responses is enhanced by differing types of instructions about the measures, varying kinds of memory cues, and differing techniques for asking questions. Also needed is development of measures of social desirability as it affects response candor. A measure of response validity might allow one to in some way "correct" responses, much as does the K-scale for certain MMPI scales[11] or might suggest the need for corroboration by biochemical measures should the measure of response validity by a patient indicate likelihood of serious response distortion.

Although much has already been accomplished, recently, there has been a growing interest in accomplishing more in the area of identifying new ways of validly measuring alcohol consumption. The use of biochemical laboratory procedures has been an area of increasingly rapid growth in the past few years. Although a few biological markers show great promise, much more has to be done to allow them to achieve the wide utility that tests for the detection of other illnesses have achieved.

In general, improvement in assay methodology and ease of use is necessary for all of the newer biological tests. Most of these are relatively sophisticated procedures for the general clinical laboratory to perform in a cost-efficient manner with the type of accuracy and precision required of a test of this importance. Of course, prior to the investment of time and resources needed to advance the technology, it must be certain that tests are of sufficient sensitivity and specificity to warrant the expenditure. Some believe that CDT is good enough to warrant greater examination in an expanded role outside of the research laboratory.

Other tests are too early in their development to support this transitional leap. Even CDT needs further elucidation of several issues, including a better definition of how much alcohol must be consumed and for how long, before the test turns positive. The variability between people in the length of time before the test returns to normal after the cessation of drinking needs to be better understood. Other issues regarding individual sensitivity to the alcohol-elevating effects on CDT is needed especially in regard to age, race, sex, and years of previous alcohol exposure. For example, it has been implied that people with previously elevated CDT levels, once they have normalized, will become abnormal again quicker if the person returns to alcohol consumption. The nature of this phenomenon needs further elucidation.

Some basic questions regarding sensitivity and especially specificity of the newer biological tests still need to be answered. This is particularly important since the tests are used away from research settings, where comparisons are typically made between well-defined treatment populations and normal controls, which have been thoroughly screened for no or minimal alcohol use. In the general clinical setting, there are more shades of gray, and the tests are held up to a more unforgiving light.

Basic technical questions for many of the tests still need to be answered, such as, what is the best substrate (serum, plasma, urine, or sweat) and what are the stability characteristics of the measured substrate prior to processing and during refrigeration or freezing. In addition, questions remain regarding the effect of other drugs, both licit and illicit, on the normal levels of these substances. Most of these issues are not insurmountable given enough directed research energy. Undoubtedly, however, the answers to some of them may limit the utility of some of the initially promising procedures.

References

[1]J. Davis, M. Dysken, W. Matuzas, and S. Nasr (1983) Some conceptual aspects of laboratory tests in depression. *J. Clin. Psychol.* **44**, 21–26.

[2]R. Baldessarini, S. Finklestein, and G. Arana (1983) The predictive power of diagnostic tests and the effect of prevalence on illness. *Arch. Gen. Psychol.* **40**, 569–573.

[3]G. Arana and D. Mossman (1988) The dexamethasone suppression test and depression. *Neurol. Clin.* **6**, 21–39.

[4]J. Swets (1986) Forms of empirical ROCS in discrimination and diagnostic tasks: Implications for theory and measurement performance. *Psychol. Bull.* **99**, 181–198.

[5]A. S. Rosman and C. S. Lieber (1990) Biochemical markers of alcohol consumption. *Alcohol Health Res.World* **14**, 210–218.

[6]M. Irwin, S. Baird, T. L. Smith, and M. Schuckit (1988) Use of laboratory tests to monitor heavy drinking by alcoholic men discharged from a drinking program. *Am. J. Psychol.* **145**, 595–599.

[7]D. G. Mayfield, G. McLeod, and P. Hall (1974) The CAGE questionnaire: Validation of a new alcoholism screening instrument. *Am. J. Psychol.* **131**, 1121–1123.

[8]M. L. Selzer, A. Vinokur, and L. Van Rooijen (1975) A self-administered Short Michigan Alcoholism Screening Test (SMAST). *J. Stud Alcohol* **36**, 117–126.

[9]R. L. Spitzer and J. B. W. Williams (1988) *Instruction manual for the structured Clinical Interview for DSM-III-R* (New York State Psychiatric Institute, New York).

[10]A. T. McLellan, L. Luborsky, G. E. Woody, and C. P. O'Brien (1980) An improved diagnostic evaluation instrument for substance abuse patients: The addiction severity index. *J. Nerv. Ment. Dis.* **168**, 26–33.

[11]S. R. Hathaway and J. C. McKinley (1989) *MMPI-2: Manual for Administration and Scoring* (University of Minnesota, Minneapolis, MN).

Index